D0295102

Novartis Foundation Symposium 255

RETINAL DYSTROPHIES:
FUNCTIONAL GENOMICS
TO GENE THERAPY

2004

**Foundation
Fighting Blindness**
Driving research to save & restore sight

John Wiley & Sons, Ltd

MEDICAL LIBRARY
QUEEN'S MEDICAL CENTRE

Copyright © Novartis Foundation 2004
Published in 2004 by John Wiley & Sons Ltd,
           The Atrium, Southern Gate,
           Chichester PO19 8SQ, UK

   National   01243 779777
   International (+44) 1243 779777
   e-mail (for orders and customer service enquiries): cs-books@wiley.co.uk
   Visit our Home Page on http://www.wileyeurope.com
              or http://www.wiley.com

All Rights Reserved. No part of this book may be reproduced, stored in a retrieval
system or transmitted in any form or by any means, electronic, mechanical, photocopying,
recording, scanning or otherwise, except under the terms of the Copyright, Designs and
Patents Act 1988 or under the terms of a licence issued by the Copyright Licensing Agency Ltd,
90 Tottenham Court Road, London W1T 4LP, UK, without the permission in writing
of the Publisher. Requests to the Publisher should be addressed to the Permissions Department,
John Wiley & Sons Ltd, The Atrium, Southern Gate, Chichester, West Sussex PO19 8SQ,
England, or emailed to permreq@wiley.co.uk, or faxed to (+44) 1243 770620.

This publication is designed to provide accurate and authoritative information in regard to
the subject matter covered. It is sold on the understanding that the Publisher is not engaged
in rendering professional services. If professional advice or other expert assistance is
required, the services of a competent professional should be sought.

Other Wiley Editorial Offices

John Wiley & Sons Inc., 111 River Street, Hoboken, NJ 07030, USA

Jossey-Bass, 989 Market Street, San Francisco, CA 94103-1741, USA

Wiley-VCH Verlag GmbH, Boschstr. 12, D-69469 Weinheim, Germany

John Wiley & Sons Australia Ltd, 33 Park Road, Milton, Queensland 4064, Australia

John Wiley & Sons (Asia) Pte Ltd, 2 Clementi Loop #02-01, Jin Xing Distripark, Singapore
129809

John Wiley & Sons Canada Ltd, 22 Worcester Road, Etobicoke, Ontario, Canada M9W 1L1

Wiley also publishes its books in a variety of electronic formats. Some content that appears
in print may not be available in electronic books.

1 0036 09 31

Novartis Foundation Symposium 255
ix+216 pages, 34 figures, 7 tables

British Library Cataloguing in Publication Data

A catalogue record for this book is available from the British Library

ISBN 0 470 85357 3

Typeset in 10½ on 12½ pt Garamond by Dobbie Typesetting Limited, Tavistock, Devon.
Printed and bound in Great Britain by T. J. International Ltd, Padstow, Cornwall.
This book is printed on acid-free paper responsibly manufactured from sustainable forestry,
in which at least two trees are planted for each one used for paper production.

Contents

Participants

Gustavo Aguirre   James A. Baker Institute of Animal Health, College of Veterinary Medicine, Cornell University, Ithaca, NY 14853, USA

Robin Ali   Department of Molecular Genetics, Institute of Ophthalmology, 11–43 Bath Street, London EC1V 9EL, UK

Wolfgang Baehr   Moran Eye Centre, University of Utah, 75 North Medical Drive, Salt Lake City, UT 84132-0001, USA

Jean Bennett   Ophthalmology, Scheie Eye Institute of Penn, 310 Stellar-Chance 422 Curie Boulevard, Philadelphia, PA 19104-6069, USA

Shomi Bhattacharya *(Chair)*   Department of Molecular Genetics, Institute of Ophthalmology, 11–43 Bath Street, London EC1V 9EL, UK

Alan C. Bird   Department of Clinical Ophthalmology, Moorfields Eye Hospital, City Road, London EC1V 2PD, UK

Dean Bok   Neurobiology & Ophthalmology, University of California, 100 Stein Plaza, Los Angeles, CA 90095-7065, USA

Hanno Bolz *(Novartis Foundation Bursar)*   Institute for Human Genetics, Univ-Krankenhaus Eppendorf, Butenfeld 42, D-22529 Hamburg, Germany

R. Kim Brazzell   Novartis Ophthalmics, 11695 Johns Creek Parkway, Duluth, GA 30097, USA

Gerald J. Chader   The Foundation Fighting Blindness, 11435 Cronhill Drive, Owings Mills, MD 21117, USA

Sheila Connelly   Genetic Therapy, Inc., 9 West Watkins Mill Road, Gaithersburg, MD 20878, USA

Frans Cremers   Department of Human Genetics, University Medical Center Nijmegen, PO Box 9101, 6500 HB Nijmegen, The Netherlands

Stephen P. Daiger   Human Genetics Center, School of Public Health, The University of Texas Health Science Center, 1200 Herman Pressler Street, Houston, TX 77030, USA

Thaddeus Dryja   Massachusetts Eye & Ear Infirmary, 243 Charles Street, Boston, MA 02114-3096, USA

Debora B. Farber   Jules Stein Eye Institute, UCLA School of Medicine, 100 Stein Plaza, Los Angeles, CA 90095-7000, USA

Martin Friedlander   Department of Cell Biology, The Scripps Research Institute, 10550 North Torrey Pines Road, MB-28, La Jolla, CA 92037, USA

Andreas Gal   Institute for Human Genetics, Univ-Krankenhaus Eppendorf, Butenfeld 42, Hamburg, D-22529, Germany

Paul Hargrave   Ophthalmology, University of Florida, Box 100284, 1600 SW Archer Road, Gainesville, FL 32610-0284, USA

William Hauswirth   Ophthalmology and Molecular Genetics, Box 100266, 1600 SW Archer Road, University of Florida, Gainesville, FL 32610-0266, USA

David M. Hunt   Division of Molecular Genetics, Institute of Ophthalmology, 11–43 Bath Street, London EC1V 9EL, UK

Michael Kaleko   Genetic Therapy, Inc., 9 West Watkins Mill Road, Gaithersburg, MD 20878, USA

George N. Lambrou   Novartis Ophthalmics AG, WKL-127.1.06, PO Box, CH-4002 Basel, Switzerland

Matthew LaVail   Beckman Vision Center, UCSF School of Medicine, 10 Kirkham Street (Room K-120), San Francisco, CA 94143-0730, USA

Roderick McInnes   Room 11101 Elm Wing, The Research Institute, Hospital for Sick Children, 555 University Avenue, Toronto, Ontario M5G 1X8, Canada

Robert S. Molday   Biochemistry/Molecular Biology, 2146 Health Sciences Mall, University of British Columbia, Vancouver, BC V6T 1Z3, Canada

Jeremy Nathans   Johns Hopkins Medical School, PCTB Room 805, 725 North Wolfe Street, Baltimore, MD 21205-2185, USA

José-Alain Sahel   Chef de Service, CHNO des Quinze-Vingts, 28 rue de Charenton, F-75557 Paris Cedex 12, France

Paul Sieving   National Institutes of Health, National Eye Institute, 31 Center Drive, Building 31 6A-03, MSC 2510, MD 20892-2510, USA

Anand Swaroop   Ophthalmology, University of MI-Kellogg Eye Center, 1000 Wall Street Room 539, Ann Arbor, MI 48105-0714, USA

Debra Thompson   Ophthalmology & Biochemistry, University of MI-Kellogg Eye Center, 1000 Wall Street Room 533, Ann Arbor, MI 48105-0714, USA

Gabriel Travis   Jules Stein Eye Institute, UCLA School of Medicine, 100 Stein Plaza, Room BH-667, Los Angeles, CA 90095, USA

Donald J. Zack   Ophthalmology, Molecular Biology and Genetics, Johns Hopkins University School of Medicine, 809 Maumenee, 600 N. Wolfe Street, Baltimore, MD 21287, USA

Chair's introduction

Shomi Bhattacharya

Department of Molecular Genetics, Institute of Ophthalmology, 11–43 Bath Street, London EC1V 9EL, UK

In introducing this symposium, I'd like to begin by looking at where we stand in relation to genes and loci associated with retinal dystrophies. Stephen Daiger has a helpful diagram charting the numbers of mapped and cloned retinal disease genes on his website, RetNet (*http://www.sph.uth.tmc.edu/RetNet/*; see Fig. 1 on page 19).

This graph shows that there are more than 135 genetic loci now mapped for retinal dystrophies, and something in the region of 80 genes that have been cloned and characterized. The types of dystrophies that we are familiar with can be broken down into three broad categories, namely retinitis pigmentosa, cone/rod dystrophies and macular diseases. We will hear in detail about the clinical aspects from Alan Bird at this meeting. The number of genes that have been associated with retinal dystrophies tells us a lot in relation to how the cell functions and its characteristics, and when there are mutations in these genes how this leads to dysfunction and cell death. How cells are actually killed by mutations is still debatable. Some of the potential therapies that are currently being developed would rely on this critical information about how the cell reacts to the mutation and eventually how it succumbs.

The key issues which face us are in the following four areas. First genetics, because practically all these conditions are inherited so we assume that there is a genetic mutation that results in the phenotype. Clearly, genetics has a major role. Then, if we know the gene, what are the functional consequences of the gene mutation? This can be understood through functional genomics. As you all know, a great deal of work is underway in this area. Third, there is the issue of animal models: when we know the gene, can we generate an animal model to mimic the disease situation? This will be extremely helpful. Of course, the development of animal models would also play an important role in the fourth topic we will be considering, which is developing gene therapy approaches for human patients. I recognize that in relation to therapy there are various possibilities that are being considered at the moment, including self transplantation, pharmacogenomics, or augmenting or removing the faulty gene in the appropriate cell type.

Under the 'heading' genetics, what are the key issues? Steve Daiger's compilation shows us 135 loci with retinal pigment genes isolated. I have no doubt that there are still many more genes to identify. Steve will be addressing this in his paper. We will be addressing the extent of genetic heterogeneity, how many more genes remain to be identified, and then the approaches for novel gene identification. The proportion of patients with mutations that have been identified is also an issue. Again, I would like to think that a significant proportion of patients with retinal dystrophies now know what mutations they have, but clearly there are many patients out there in whom the gene hasn't been identified and the mutation is not known.

The human genome has been more-or-less sequenced, and a large number of novel genes have been identified. We need to know whether a proportion of these genes are expressed in the eye in the appropriate cell types in the retina: some could be appropriate candidates for these types of retinal dystrophies. The contribution of the human genome project to our endeavour needs to be elucidated.

Bioinformatics is a major issue. Now we can do a lot of gene identification and characterization *in silico*. This will play a key role in developing our understanding of the retinal dystrophy genes and mutations.

For clinicians and scientists involved in this endeavour, the understanding of the interrelationship of genotype and phenotype will be a key issue. Given the extent of heterogeneity seen in patients, it is important to be able to correlate phenotype and genotype, which will help improve the clinical diagnosis and management of patients.

In functional genomics we will discuss the identification of novel genes with novel function. The biochemical basis of disease will be discussed later when David Hunt talks about the retGC/GCAP mutation and its role in how the disease might develop. Of course, as we know more about the genes we will understand the biochemical basis of the disease, and this is a key issue. Then on the subject of the progression from genes to cellular pathways there will be papers by Anand Swaroop and Don Zack. These are based on a functional genomics approach. Once again, the key issue in our understanding may develop from this kind of functional genomics work, looking at the progression from genes to cellular pathways. Microarrays will be one of the key items in our discussions. Don Zack and Anand Swaroop will both present data on these. Will observed alterations in gene expression help in our understanding of the disease pathways? I have no doubt that microarray analyses will make a major contribution, telling us a great deal about the disease biology. This same technology will also enable us to identify suitable candidate genes. The most recent success is the cloning of the dominant RP locus on chromosome 7q. This was on the basis of microarray SAGE analysis where candidate genes have been identified and one of these genes mapped in the same locus as the 7q dominant RP locus.

A key issue regarding animal models is how we can generate them. Do they truly represent human diseases, and will our understanding of the disease process come from studying these models? Is an animal model necessary for each gene implicated in retinal degeneration? Are there going to be some generic lessons that will help in developing therapy or understanding the disease process?

The final area is gene therapy. There are other methods that are being developed for therapy, but for this meeting we are concentrating on gene therapy. The key issues here are the choice of vectors, pathogenicity of the vectors and the delivery system. Also, are large animal trials necessary before clinical trials can begin? In this regard we are going to hear a talk from Jean Bennett who will talk about the RP65 dog model. There are clear issues about efficacy and safety. Another question concerns whether some diseases are more amenable to gene therapy. Which disease gene should be the first target for gene therapy? And what assessment criteria should we use? We need to know whether we have made some objective improvement in the patient, so good criteria are needed for evaluating therapy. Finally, and very importantly, there are issues of consent and ethical approval.

These are the challenges that I see facing us during this meeting. I hope that we will find answers to many of these questions and raise other important issues for future investigation.

Gene therapy of retinal dystrophies: achievements, challenges and prospects

Dean Bok

Department of Neurobiology, Jules Stein Eye Institute and Brain Research Institute, University of California, Los Angeles, CA 90095, USA

Abstract. Early attempts at gene therapy of inherited retinal diseases by recombinant adenovirus-vectored gene replacement in laboratory animals met with moderate success but the effect was transient. Recently, emphasis has shifted to less toxic vectors, namely recombinant adeno-associated (rAAV) viruses. Ribozymes, targeted to the P23H rhodopsin mutation in transgenic rats, significantly reduced photoreceptor loss and slowed attenuation of the electroretinogram (ERG) for 8 months. By gene replacement, rAAV-based photoreceptor rescue has been achieved in the $rds^{-/-}$ mouse and has restored vision in dogs carrying a RPE65 gene mutation. Minigenes for neurotrophins delivered by rAAV have been effective in achieving structural rescue of photoreceptors in rodent models of dominant disease, although this has not always been accompanied by functional rescue. One of the current challenges is the application of ribozyme therapy for dominant mutations coupled with wild-type gene augmentation to overcome haploinsufficiency. Other animal models are currently being utilized for preclinical studies as well. Spontaneously mutated Irish Setters and *rd* mice offer excellent subjects for the therapy of recessive mutations as do the RPE65 knockout mouse and RCS (*rdy*) rat. With burgeoning preclinical successes, the future looks bright for the treatment and cure of inherited retinal diseases in human patients.

2004 Retinal dystrophies: functional genomics to gene therapy. Wiley, Chichester (Novartis Foundation Symposium 255) p 4–16

The road from gene discovery to gene therapy for inherited retinal diseases has taken an interesting path. In 1984, Bhattacharya and collaborators were the first to report a gene linkage for a non-syndromic inherited retinal disease, specifically an X-linked form of retinitis pigmentosa (RP) (Bhattacharya et al 1984). However, due to technical difficulties the identity of a gene underlying this inherited class of RP was not the first to be solved. Instead, a RP linkage on the long arm of chromosome 3, reported by the laboratories of Humphries and Daiger (McWilliam et al 1989) was the first to associate RP with a characterized gene product. To the surprise of many this gene was *rhodopsin* (Dryja et al 1990).

4

When first informed of this important discovery, it took this author and his contemporaries back to a meeting co-sponsored by the Retinitis Pigmentosa (RP) Foundation (now the Foundation Fighting Blindness) in the early 1970s. The Canadian and American RP Foundations sponsored a meeting of basic scientists and clinicians at a country resort called The Briars near Toronto. At that period in the history of the field, there were no credible insights into the molecular aetiology of RP. Although there was evidence for a genetic basis, there were multiple, conflicting theories based on vascular attenuation, inflammation, lysosomal leakage within the retinal pigment epithelium (RPE) and others. At the conference, I recall that a young, brilliant investigator named Richard Cone (an appropriate name for a vision scientist) offered his opinion that rhodopsin must somehow be involved in RP. Perhaps he was inspired by the burgeoning interest in phototransduction or maybe he was truly clairvoyant. In any event, his prediction turned out to be true.

The path to discovery of rhodopsin as an important player in RP was an arduous one and this success story is worth repeating because it points out the importance of fundamental science in any endeavour that involves the conquest of human disease. Following the development of methods for the purification of rhodopsin, it required approximately 15 years of hard work before the primary structure of this intrinsic membrane protein was solved (Ovchinnikov et al 1983, Hargrave et al 1983). This crucial information was then utilized by a gifted MD/PhD student named Jeremy Nathans at Stanford University who determined the structure of the human rhodopsin gene (Nathans & Hogness 1984). This information was essential for the screening of human DNA for the first disease-causing rhodopsin gene allele.

Thus, the prospect for gene-based approaches toward the treatment of inherited retinal diseases became a reality and we are gathered here to assess the past, present and future. Gene therapy to date has enjoyed limited success in non-ocular tissues but is now coming into prominence in treating the retina of the eye. One of the problems encountered in gene therapy of non-ocular tissues has been the premature application of this approach in human patients without sufficient investigator awareness of pitfalls. All of the work involving retinal gene therapy to date has been performed in preclinical settings (animal models) and this has been highly beneficial to our field.

Practitioners of the art and science of gene therapy in the eye have had the benefit of learning from initial setbacks in other fields, the development of improved virus vectors and the relatively favourable setting in which these vectors are placed into the eye, namely the 'subretinal space'. This region is not a true space but rather an extracellular matrix-rich compartment that separates the two major components of the retina. These are the neurosensory portion, which includes the photoreceptors, the cells directly involved in initiation of the visual response, and the RPE—a cellular monolayer that nourishes the photoreceptors. Incarceration of the virus vectors within this space promotes interaction with the target cells, usually the photoreceptors or the RPE, and allows for the use of relatively low doses but high local concentrations around the cells that are to be transduced by the virus. Thus, no sophisticated homing systems are required.

The examples of gene therapy that will be used to support these statements are all taken from preclinical work, as are most of the proposals for future work. Nonetheless, Phase I/II clinical trials may be anticipated within a few years.

Gene replacement and ribozyme strategies

During the past five years, proof of principle for gene-based therapy of inherited retinal diseases has been established in animal models. Initial studies utilized first-generation recombinant adenoviral (rAV) vectors (Bennett et al 1996) and demonstrated moderate, but transient success. The results were compromised in part by the complexity of the vector genome and its products, and significant stimulation of the immune response to the point that expression of the rescue gene was suppressed (Bennett et al 1996). Even when a minimal ('gutted') rAV was used as a vector, transient rescue gene expression was still observed (Kumar-Singh & Farber 1998). Considerably more promising results have been reported with recombinant adeno-associated (rAAV) virus which, when used as a vector is stripped of all of its genes with the exception of short viral inverted terminal repeats (Hauswirth et al 2000). The parent virus, which is non-pathogenic and stably integrates into an indifferent site in the human genome produces circulating antibodies in its host (about 80% of humans have antibodies to this virus) and the rAAV also produces a mild immune response (Bennett et al 1999). However several laboratories have reported long-term stable expression of genes delivered to the retina in rAAV (Bennett et al 1999, Hauswirth et al 2000), even following a second treatment in the presence of circulating antibodies to rAAV capsid proteins (Bennett et al 1999).

The first, striking success with rAAV in an animal model for inherited retinal degeneration was obtained with a ribozyme strategy (Lewin et al 1998). These investigators used transgenic rats carrying the P23H opsin gene mutation, which causes retinitis pigmentosa in humans and is dominant in its action. Coding sequences for hairpin and hammerhead ribozymes were designed to cleave a specific site in the coding sequence of the mutant mRNA, leaving the wild type mRNA intact. These ribozyme coding sequences were then placed under control of the bovine opsin promoter to confer cell-specific expression in rods and packaged into rAAV. Following injection into the subretinal space there was significant slowing of the rate of photoreceptor degeneration over a period of 3 months. Subsequent studies now indicate significant, persistent rescue for at least 8 months (LaVail et al 2000). In addition to morphological rescue, physiological rescue was also observed as measured by comparative ERGs in the injected and contralateral non-injected eye. Hence, there is considerable potential for this strategy in the treatment of dominantly inherited disease where the dominant-negative effect of the mutant allele must first be removed before replacement gene therapy can be effective. In the case of dominant-negative rhodopsin mutations, replacement of the mutant allele mRNA would probably be necessary as well. It has been shown that disruption of one allele in these mice allows the formation of normal rod outer segments initially but results in slow photoreceptor degeneration over time (Lem et al 1999). Replacement would be even more important for rds/peripherin, whose gene displays haploinsufficiency, as described below.

Others (Ali et al 2000) have recently shown that a replacement minigene vectored by rAAV can achieve partial rescue in $rds^{-/-}$ mice. This is encouraging, additional evidence for the efficacy of the rAAV strategy. Ali and colleagues were able to elicit the formation of truncated outer segments in $rds^{-/-}$, which carries a null, insertion mutation in this gene (Travis et al 1989). Rds/peripherin, the protein product of the rds gene (Connell et al 1991, Travis et al 1991) is a member of a small family of adhesion molecules that are essential for the formation of rod and cone photoreceptor outer segment discs. In its absence, no discs are formed (Nir & Papermaster 1986, Usukura & Bok 1987). Heterozygous ($rds^{+/-}$) mice, with a spontaneous null mutation in one allele, have oversized outer segment discs that curl into whorls (Hawkins et al 1985). Thus the gene manifests haploinsufficiency (semidominance). The gene therapy strategy used by Ali et al (2000) was quite successful in $rds^{-/-}$ mice and would presumably be even more so in $rds^{+/-}$. However, this is not the genotype for Rds/peripherin-based disease in the human population, which has over 60 different mutations reported thus far. Most human patients have dominant-negative point mutations (Sohocki et al 2001) rather than null mutations, which would require a combination of targeted ribozyme plus rds gene replacement therapy. Fortunately, all of this could be achieved in the context of an AAV vector and this work is in progress in our laboratory in collaboration with William Hauswirth, Alfred Lewin and Matthew LaVail. To that end, we are using a transgenic mouse model with a P216L, dominant-negative point mutation in Rds/peripherin (Kedzierski et al 1997) as the experimental subject.

Perhaps the most dramatic success with retinal gene therapy to date, at least in the context of its emotional impact, has been the recent work of Acland et al (2001) on the Briard dog. Some of the animals in this breed carry a mutation in the RPE65 gene. Previous studies on mice with a disrupted RPE65 gene by Redmond et al (1998), revealed that the protein product of this gene is essential for the production of 11-cis-retinol, the immediate precursor for 11-cis-retinal. The latter is the essential rod and cone photoreceptor opsin chromophore that initiates vision. It was subsequently discovered that a mutation in this gene is the cause of the very early onset vision defect in these dogs (Aguirre et al 1998). Also, it was known at that time that this gene is the basis for a form of Leber congenital amaurosis (blindness) in humans (Marlhens et al 1997, Gu et al 1997). Acland et al (2001) using rAAV vector-based gene therapy, reported behavioural and electrophysiological evidence for the acquisition of visual function in young Briard dogs that were sightless prior to therapy. The prospect of treating young children in the same manner and offering them sight is most heart-warming indeed. This is one of two examples of success in the treatment of an inherited retinal disease in which the primary site of expression of the mutant gene is the RPE.

The second example, a very recent one, involving treatment of an RPE-specific disease comes from the venerable RCS (*rdy*) rat. This was one of the first putative animal models for human retinitis pigmentosa, but one whose causative gene had been a mystery for over 30 years, until D'Cruz et al (2000) solved this elusive riddle. They found that the defective gene is a receptor tyrosine kinase called Mertk, an RPE membrane receptor apparently essential for the transmembrane signalling event during the daily phagocytosis of photoreceptor outer segment fragments. Hard on the heels of this important discovery was the detection of gene defects in humans as well (Gal et al 2000) finally, establishing the RCS rat as a *bona fide* animal model for retinitis pigmentosa. Vollrath et al (2001) have now partially corrected the retinal dystrophy phenotype in RCS rats by adenovirus-based gene transfer. Transduction of the RCS RPE with wild-type Mertk reversed the phagocytic defect, elicited considerable rescue of photoreceptors from death and improved the cornea-negative scotopic threshold response by two log units in the treated eyes.

The recent work of Acland et al (2001) and Vollrath et al (2001) provides compelling evidence that RPE-based inherited diseases can be treated as readily as those that are photoreceptor based. Interestingly, in both cases, correction of an expressed gene defect in the RPE was able to correct a photoreceptor phenotype. These phenotypes are characterized by absence of rhodopsin regeneration due to lack of 11-*cis*-retinal chromophore and slow photoreceptor degeneration in the case of the Briard dog and photoreceptor dystrophy and cell death in the RCS rat.

Gene-based trophic factor therapy

The seminal work of LaVail and Steinberg et al (Faktorovich et al 1990, LaVail et al 1992) demonstrated that a variety of growth factors and cytokines, which include leukaemia inhibitory factor (LIF) and ciliary neurotrophic factor (CNTF), rescue rodent photoreceptors from genetic and environmental insult for up to a month when injected as a single bolus into the vitreous cavity of the rat eye. It is not known how these factors exert their rescue effect on photoreceptors. To date, there is no evidence for CNTF receptors on adult rodent photoreceptors nor are their photoreceptors known to express CNTF. The nearest neighbour in which CNTF protein and mRNA have been detected is the Müller glial cell (Kirsch et al 1997). However, Beltran et al (2002) have recently reported expression of the CNTFRα subunit in canine photoreceptors.

Prompted by the observation that CNTF exerts a rescue effect on adult photoreceptors, which do not express CNTF protein or the neuron-specific subunit CNTFRα, several investigators have sought to determine whether the effect of CNTF is direct or indirect by studying CNTF-mediated signalling pathways in retinal neurons and glia. CNTF activates the JAK (Janus tyrosine kinase) and STAT (signal transducers and activators of transcription) signalling pathway (Stahl & Yoncopoulos 1994), hence investigators have looked for up-regulation of mRNAs for these proteins as a function of CNTF stimulation. CNTF can also activate ERK1 and ERK2 (Boulton et al 1994) so there is some cross talk for CNTF. Wahlin et al (2000) gave single injections of various factors (BDNF, CNTF and FGF2) into the mouse vitreous and observed a rapid increase in phosphorylated ERK (pERK) in Müller cells and an increase in c-*fos* in Müller, amacrine and ganglion cells. These levels returned to baseline by 6–24 h but were followed by increases in Müller cell GFAP. Peterson et al (2000), using rats, injected Axokine™ (a recombinant form of CNTF) into the vitreous and observed phosphorylation of STAT3 (pSTAT3) and ras-MAPK. pSTAT3 was localized to nuclei of retinal Müller cells, ganglion cells and astrocytes but not photoreceptors. These data, taken together suggest an indirect effect of CNTF on photoreceptor survival in rodents.

When LaVail and associates attempted to extend their studies on injected CNTF rescue to various mouse models for RP, they achieved only limited success and had none with *rds*$^{-/-}$ mice (LaVail et al 1998). They suggested that better methods of CNTF delivery might be indicated. Supporting this hypothesis, Cayouette et al (1998), using intravitreal injections of an adenovirus-vectored, secreted form of CNTF, observed transient photoreceptor rescue in *rds*$^{-/-}$ mice. They observed an improvement in the scotopic and photopic ERG and, surprisingly, enhanced opsin expression, in spite of the fact that these null mice cannot express Rds/peripherin. We have recently had success in rescuing the photoreceptors of transgenic mice that carry a Rds/peripherin P216L point mutation with AAV-vectored, secreted CNTF (Bok et al 2002). However, our result, which was not transient in terms of morphological rescue, was also very different with respect to the ERG. An unexpected result was suppression of the scotopic a and b-wave and photopic b-wave in the injected eye compared to the non-injected, contralateral eye, even though there was significant morphological rescue of rods in the injected eye as judged by the thickness of the outer nuclear layer. Additionally, there was a striking change in photoreceptor nuclear morphology from the condensed chromatin pattern typical of rods to a more diffuse pattern reminiscent of cones of bipolar cells. Liang et al (2001) reported similar aberrations when using rAAV-vectored CNTF in transgenic rats carrying P23H and S334ter mutations whereas Lau et al (2000) reported photoreceptor rescue in tandem with ERG rescue when S334ter rats were treated with rAAV-mediated fibroblast growth factor 2.

Conclusions

Gene therapy of retinal dystrophies has enjoyed extraordinary progress since the first encouraging preclinical study was reported just seven years ago. It is likely that the first human safety trials will be conducted within the next five years and it is essential that the most favourable gene or genes be chosen for these initial trials. Failures with gene therapy are well known in other fields and ours can ill afford such a setback. This meeting should include a discussion of the genes that are most likely to be amenable to successful transfer into human patients.

References

Acland GM, Aguirre GD, Ray J et al 2001 Gene therapy restores vision in a canine model of childhood blindness. Nat Genet 28:92–95

Aguirre GD, Baldwin V, Pearce-Kelling S, Narfstrøm K, Ray K, Acland GM 1998 Congenital stationary night blindness in the dog: common mutation in the RPE65 gene indicates founder effect. Mol Vis 4:23

Ali RR, Sarra G-M, Stephens C et al 2000 Restoration of photoreceptor ultrastructure and function in retinal degeneration slow mice by gene therapy. Nat Genet 25:306–310

Bhattacharya SS, Wright AF, Clayton JF et al 1984 Close genetic linkage between X-linked retinitis pigmentosa and a restriction fragment length polymorphism identified by recombinant DNA probe L1.28. Nature 309:253–255

Beltran WA, Kijas JW, Zhang Q, Aguirre GD 2002 Cloning and expression of ciliary neurotrophic factor receptor (CNTFRα) in the canine retina. ARVO (abstr 2679)

Bennett J, Tanabe T, Sun D et al 1996 Photoreceptor cell rescue in retinal degeneration (rd) mice by in vivo gene therapy. Nat Med 2:649–654

Bennett J, Maguire AM, Cideciyan AV et al 1999 Stable transgene expression in rod photoreceptors after recombinant adeno-associated virus-mediated gene transfer to monkey retina. Proc Natl Acad Sci USA 96:9920–9925

Bok D, Yasumura D, Matthes MT et al 2002 Effects of adeno-associated virus-vectored ciliary neurotrophic factor on retinal structure and function in mice with a P216L rds/peripherin mutation. Exp Eye Res 74:719–735

Boulton TG, Stahl N, Yancopoulos GD 1994 Ciliary neurotrophic factor/leukemia inhibitory factor, interleukin6/oncostatin M family of cytokines induces tyrosine phosphorylation of a common set of proteins overlapping those induced by other cytokines and growth factors. J Biol Chem 269:11648–11655

Cayouette M, Behn D, Sendtner M, Lachapelle P, Gravel C 1998 Intraocular gene transfer of ciliary neurotrophic factor prevents death and increases responsiveness of rod photoreceptors in the retinal degeneration slow mouse. J Neurosci 18:9282–9293

Connell G, Bascom R, Molday L, Reid D, McInnes RR Molday RS 1991 Photoreceptor peripherin is the normal product of the gene responsible for retinal degeneration in the *rds* mouse. Proc Natl Acad Sci USA 88:723–726

D'Cruz PM, Yasumura D, Weir 2000 Mutation of the receptor tyrosine kinase gene Mertk in the retinal dystrophic RCS rat. Hum Mol Genet 9:645–651

Dryja TP, McGee TL, Hahn LB et al 1990 Mutations within the rhodopsin gene in patients with autosomal dominant retinitis pigmentosa. N Engl J Med 323:1302–1307

Faktorovich EG, Steinberg RH, Yasumura D, Matthes MT, LaVail MM 1990 Photoreceptor degeneration in inherited retinal dystrophy delayed by basic fibroblast growth factor. Nature 347:83–86

Gal A, Li Y, Thompson DA et al 2000 Mutations in MERTK, the human orthologue of the RCS rat retinal dystrophy gene, cause retinitis pigmentosa. Nat Genet 26:270–271

Gu SM, Thompson DA, Srikumari CR et al 1997 Mutations in RPE65 cause autosomal recessive childhood-onset severe retinal dystrophy. Nat Genet 17:194–197

Hargrave PA, McDowell JH, Curtis DR et al 1983 The structure of bovine rhodopsin. Biophys Struct Mech 9:235–244

Hauswirth WW, LaVail MM, Flannery JG, Lewin AS 2000 Ribozyme gene therapy for autosomal dominant retinal disease. Clin Chem Lab Med 38:147–153

Hawkins RK, Jansen HG, Sanyal S 1985 Development and degeneration of retinal in *rds* mutant mice: photoreceptor abnormalities in the heterozygotes. Exp Eye Res 41:701–720

Kedzierski W, Lloyd M, Birch DG, Bok D, Travis GH 1997 Generation and analysis of transgenic mice expressing P216L-substituted rds/peripherin in rod photoreceptors. Invest Ophthalmol Vis Sci 38:498–509

Kirsch M, Lee M-Y, Meyer V, Wiese A, Hofmann H-D 1997 Evidence for multiple, local functions of ciliary neurotrophic factor (CNTF) in retinal development: expression of CNTF and its receptor and in vitro effects on target cells. J Neurochem 68:979–990

Kumar-Singh R, Farber DB 1998 Encapsidated adenovirus mini-chromosome-mediated delivery of genes to the retina: application to the rescue of photoreceptor degeneration. Hum Mol Genet 7:1893–1900

Lau D, McGee LH, Zhou S et al 2000 Retinal degeneration is slowed in transgenic rats by AAV-mediated delivery of FGF-2. Invest Ophthalmol Vis Sci 41:3622–3633

Lem J, Krasnoperova NV, Calvert PD 1999 Morphological, physiological and biochemical changes in rhodopsin knockout mice. Proc Natl Acad Sci USA 96:736–741

LaVail MM, Unoki K, Yasumura D, Matthes MT, Yancopoulos GD, Steinberg RH 1992 Multiple growth factors, cytokines and neurotrophins rescue photoreceptors from the damaging effects of constant light. Proc Natl Acad Sci USA 89:11249–11253

LaVail MM, Yasumura D, Matthes MT et al 1998 Protection of mouse photoreceptors by survival factors in retinal degenerations. Invest Ophthalmol Vis Sci 39:592–602

LaVail MM, Yasumura D, Matthes MT et al 2000 Ribozyme rescue of photoreceptor cells in P23H transgenic rats: long-term survival and late-stage therapy. Proc Natl Acad Sci USA 97:11488–11493

Lewin AS, Drenser KA, Hauswirth WW 1998 Ribozyme rescue of photoreceptor cells in a transgenic rat model of autosomal dominant retinitis pigmentosa. Nat Med 4:967–971

Liang F-Q, Aleman TS, Dejneka NS et al 2001 Long-term protection of retinal structure but not function using rAAV.CNTF in animal models of retinitis pigmentosa. Mol Ther 4:461–472

Marlhens F, Bareil C, Griffoin JM et al 1997 Mutations in RPE65 cause Leber's congenital amaurosis. Nat Genet 17:139–141

McWilliam P, Farrar G J, Kenna P et al 1989 Autosomal dominant retinitis pigmentosa (ADRP): localization of an ADRP gene to the long arm of chromosome 3. Genomics 5:619–622

Nathans J, Hogness DS 1984 Isolation and nucleotide sequence of the gene encoding human rhodopsin. Proc Natl Acad Sci USA 81:4851–4855

Nir I, Papermaster D 1986 Opsin gene expression during early and late phases of retinal degeneration in *rds* mice. Exp Eye Res 51:257–267

Ovchinnikov YA, Abdulaev NG, Feigina MY et al 1982 The complete amino acid sequence of visual rhodopsin. Bioorg Khim 8:1011–1014

Peterson WM, Wang Q, Tzekova R, Wiegand SJ 2000 Ciliary neurotrophic factor and stress stimuli activate the Jak-STAT pathway in retinal neurons and glia. J Neurosci 20:4081–4090

Redmond TM, Yu S, Lee E et al 1998 Rpe65 is necessary for production of 11-cis Vitamin A in the retinal visual cycle. Nat Genet 20:344–351

Sohocki MM, Daiger SP, Bowne SJ et al 2001 Prevalence of mutations causing retinitis pigmentosa and other inherited retinopathies. Hum Mutat 17:42–51

Stahl N, Yancopoulos GD 1994 The tripartite CNTF receptor complex: activation and signaling involves components shared with other cytokines. J Neurobiol 25:1454–1466

Travis GH, Brennan MB, Danielson PE, Kozak CA, Sutcliffe JG 1989 Identification of a photoreceptor-specific mRNA encoded by the gene responsible for retinal degeneration slow (*rds*). Nature 338:70–73

Travis GH, Sutcliffe JG, Bok D 1991 The retinal degeneration slow (*rds*) gene product is a photoreceptor disc membrane-associated glycoprotein. Neuron 6:61–70

Usukura J, Bok D 1987 Changes in the localization and content of opsin during retinal development in the *rds* mutant mouse: immunocytochemistry and immunoassay. Exp Eye Res 45:501–515

Vollrath D, Feng W, Duncan JL et al 2001 Correction of the retinal dystrophy phenotype of the RCS rat by viral gene transfer of Mertk. Proc Natl Acad Sci USA 98:12584–12589

Wahlin KJ, Campochiaro PA, Zack DJ, Adler R 2000 Neurotrophic factors cause activation of intracellular signaling pathways in Müller cells and other cells of the inner retina, but not photoreceptors. Invest Ophthalmol Vis Sci 41:927–936

DISCUSSION

[*Note added in press by William Hauswirth:* Recently, the issue of vector AAV chromosomal integration has arisen as a potential source of tumorigenesis in light of recent reports of retrovirus vector integration-related leukaemias in three of 10 treated X-linked SCIDS patients. First, it is important to emphasize that the biology of integration is very different between AAV and retroviral vectors, and that it is a mistake to draw an analogy between the two. Retroviruses lead to passenger gene expression through an obligate integration event; hence, 100% chromosomal integration is found. Additionally, retroviruses integrate relatively site specifically near a variety of pro- or anti-tumour genes, as they did in the XSCIDS cases. In contrast, AAV vectors from which effective passenger gene expression emanates are very rarely integrated (less than 0.5% of vector genomes, Schnepp et al 2003). This recent study confirms many previous, but less quantitative studies, also finding very low frequencies of integration and large amounts non-integrated vector from which expression originated (labs of Flotte, Englehardt, Xiao, Samulski). Even when the experiment is purposely biased to favour detection of integrated AAV vector, integration events were still rare (Check 2003). Finally, in over 100 patients treated to date with AAV vectors over the past 5+ years, no vector related tumours have been reported. Clearly, caution dictates that the tumour incidence in preclinical and clinical AAV gene therapy trials continues to be carefully monitored, however it is a mistake, and perhaps a disservice to the patient population, to make simple extrapolations from one vector type to another without a clear understanding of the biological differences between them.]

McInnes: The retrovirus promoters are very powerful and have distal effects. Do the promoters present in AAV have the same potential risk?

Ali: They are much weaker.

Hauswirth: This is probably because of the inverted terminal repeats that can act as silencers and/or insulators for transcription either coming into the region of going out of the region.

Ali: It may be fair to speculate that retroviral transduction of a proliferating population of cells is more likely to lead to oncogenic events than transduction of non-dividing tissue. It has been suggested that the parts of the genome that are available for integration are composed of DNaseI-sensitive sites, which implies that these are genes that are being expressed. If we think about the stem cells that were being targeted, the sorts of genes that are being expressed are possibly genes involved in cell division and cell regulation. This may not be the case in neuronal cells and photoreceptors. I agree with Dean Bok that issues surrounding integration are important — we need to understand this and be rigorous about it — but I think there may be good reasons for thinking that integration events in the retina may not present the same problem as those in proliferating stem cells.

Hauswirth: As Dean implied, if you add back the viral *rep* gene, which is the gene that mediates site-specific integration in chromosome 19 in the wild-type, this halves the amount of space available for your own gene. This leaves just 2.2 kb, which is not enough for many of the things we want to do. It has its own problems. One idea that gets us up to 5 kb of payload in the vector is to put in two viruses, one expressing the *rep* gene and one expressing the therapeutic gene. Presumably *rep* will act to target the therapeutic gene to chromosome 19, which we know is a safe site since 60% of the world's population has AAV integrated into this site.

LaVail: Can you get the titres high enough so that you can be sure that you have a doubly infected cell?

Hauswirth: This is probably not a problem. A back of the envelope calculation would suggest we have done this in several inducible situations. With a Dox-inducible system the multiplicity of infections is typically over 1000 per photoreceptor cell, so the odds of getting two vectors into one cell are very high.

McInnes: Might the choice of CNTF as the first neurotrophic factor to use in therapy retrospectively turn out to be unfortunate? Perhaps the effects it is having on the cell might not occur with other factors.

LaVail: That is a possibility. On the other hand, it is very hard to turn your back on a molecule that shows a positive rescue effect in over 13 different forms of inherited retinal degeneration. It needs to be given every chance to work. We need to see what these negative effects have on the structure of the cell, and the ramifications of this for vision over the long term. Some people have suggested it may be that the lowered amplitude of the output of the cell is actually an attempt by the photoreceptor cell to preserve its energy level and live longer under unusual circumstances. One of the points you make is that even though there is a finite number of retinal vision researchers, we do need to look at some of the other neurotrophic factors. There is, however, clearly a specificity here. At first, when people began using CNTF with better delivery systems as opposed to bolus injections, it appeared that CNTF provided rescue in every case. Thus, it appeared that all our negative findings were artifacts of the bolus injection. Does every agent work in every degeneration? I think the answer is no, on the basis of Don Zack's and Peter Campochiaro's work in transgenic animals (Yamada et al 2001) and some of our studies (M. M. LaVail, unpublished results). Certain agents need to be looked at more closely than they have been.

McInnes: It is clear that these cells might be cycling. Has anyone done labelling?

LaVail: We have never seen mitotic figures, but the photoreceptor cells are clearly more immature. There are studies showing that CNTF can impair differentiation and inhibit opsin expression.

Bok: We have hammered these cells as hard as we can. When Bill Hauswirth engineered this CNTF vector that we employed in our study, he even introduced a couple of amino acid changes that increased the affinity of the CNTF for the α subunit of its receptor. We are now cutting back on the dose to see whether we can maintain the rescue effect but get rid of the so-called 'side effects'. We are turning on a lot of different genes in these cells and this somehow impacts on the phototransduction cascade in ways that we don't understand.

Zack: Rod McInnes mentioned that one approach is to look at other factors. But there's another approach with CNTF: how about looking at cells other than photoreceptors? CNTF acts on a number of other neuronal cells. Have people studying these other neuronal cells seen evidence for change in gene expression with CNTF?

Bok: Looking at the retinas we did not perceive changes in other retinal neurons. But we really haven't explored this.

Zack: In spinal cord injury, is there evidence that CNTF is changing the neuronal cells or is it just leading to their better survival?

LaVail: In general, I don't think the other fields are nearly as far along as we are. For them it isn't as easy to look at the tissues, and they don't know exactly where they are injecting.

Farber: Adriana Di Polo has done beautiful work with CNTF on ganglion cells and she gets great rescue.

Zack: But has she looked at the function of the ganglion cells? The rescue is the first part, but it is the function that we need to maintain.

Aguirre: With regard to CNTF and the changes in the photoreceptor cells, we found marked reactive changes in the dog retina following FGF2 and FGF18 administered with AAV vectors. Although there was dramatic rescue effect in terms of the number of rows of outer nuclear layer (ONL) nuclei remaining, in some areas there were more nuclei when we ended than the number present at the time of injection. This suggested that there was an increased proliferation. Unlike the CNTF experiments in the *rcd1* retinas where rescue was dramatic and the structural preservation was quite good, the rescue was accompanied by marked abnormalities. Personally, I am not that concerned that there is a decrease in the b-wave following CNTF administration. Some of the patients with Duchenne's muscular dystrophy have a b-wave response that is changed towards a negative type ERG, but they don't report any changes in night vision defects. If you have a very positive rescue effect and there is a slight decrease in b-wave amplitude, this might not be clinically significant.

Swaroop: What do we know about CNTF's downstream pathways? Would it be an option to use some of these downstream molecules as targets for future drug discovery? What is known about CNTF function in the retina? Can we modulate downstream signalling molecule(s) that would still have the beneficial effects of CNTF with fewer side effects?

LaVail: Some people have been looking at that, but it is a little too early to make any conclusions. One of the problems is that there is a dose-dependent side effect of CNTF just as there is a dose-dependent rescue effect. We need to know about this for every single aspect of the rescue process.

Farber: Dean Bok, you talk about all these retinal degenerations in mice, rats and dogs. What has been done in this respect with primates?

Bok: There are very few studies of this type in primates, because we don't have the models. But we can now more readily create these models using knock-down experiments with vectors or by introducing mutations. There is a colony of monkeys that Paul Sieving might be interested in trying to bring to the USA from Japan. They have what seems to be a macular degeneration.

Zack: There is an interesting reverse parallel. In the glaucoma field they have a very good monkey model but limited rodent models. We are blessed with good rodent models but we are struggling for a monkey one.

References

Check E 2003 Harmful potential of viral vectors fuels doubts over gene therapy. Nature 423:573–574

Schnepp BC, Clark KR, Klemanski DL, Pacak CA, Johnson PR 2003 Genetic fate of recombinant adeno-associated virus vector genomes in muscle. J Virol 77: 3495–3504

Yamada H, Yamada E, Ando A et al 2001 Fibroblast growth factor-2 decreases hyperoxia-induced photoreceptor cell death in mice. Am J Pathol 159:1113–1120

Identifying retinal disease genes: how far have we come, how far do we have to go?

Stephen P. Daiger

Human Genetics Center, School of Public Health, and Department of Ophthalmology and Visual Science, The University of Texas Health Science Center, Houston, TX 77030, USA

Abstract. One of the great success stories in retinal disease (RD) research in the past decade has been identification of many of the genes and mutations causing inherited retinal degeneration. To date, more than 133 RD genes have been identified, encompassing many disorders such as retinitis pigmentosa, Leber congenital amaurosis, Usher syndrome and macular dystrophy. The most striking outcome of these findings is the exceptional heterogeneity involved: dozens of disease-causing mutations have been detected in most RD genes; mutations in many different genes can cause the same disease; and different mutations in the same gene may cause different diseases. Superimposed on this genetic heterogeneity is substantial clinical variability, even among family members with the same mutation. The RD genes involve many different pathways, and expression ranges from very limited (e.g. expressed in rod photoreceptors only) to ubiquitous. These findings raise several general questions — in addition to the extraordinary number of specific, biological problems revealed. What fraction of the patient population can now be accounted for by the known RD genes? How many more RD genes will be found, and how should we find them? Are we dealing with just a handful of disease mechanisms or are there many different routes to retinal degeneration? How will this extreme heterogeneity affect our ability to diagnose and treat patients? These questions are considered in this summary.

2004 Retinal dystrophies: functional genomics to gene therapy. Wiley, Chichester (Novartis Foundation Symposium 255) p 17–36

During the past 15 years many of the genes and mutations causing inherited retinal diseases (RD genes) have been identified by a variety of methods. It is not a coincidence that this progress has paralleled completion of the human genome project: many of the technical advances in the genome project have enriched RD research. Also, identification of a complete set of human genes provides a powerful research tool in all areas of medicine, not just RD research. But RD research has been exceptionally productive, and might serve as a model for other types of inherited disease with both hopeful and disturbing implications.

The hopeful aspect is the demonstrated efficacy of what are now standard methods in medical genetics for gene identification, that is, ascertainment of patients and families, linkage mapping, positional candidate cloning and candidate gene screening. The disturbing aspect is just how complex retinal diseases have proven. The complexity includes:

- *Genetic heterogeneity*: mutations in different genes may cause the same retinal disease
- *Allelic heterogeneity*: many different disease-causing mutations are found in most RD genes
- *Phenotypic heterogeneity*: different mutations within the same gene may produce different clinical phenotypes and
- *Clinical heterogeneity*: the same mutation in different individuals, even within the same family, may produce different clinical consequences.

Whether these forms of heterogeneity, which are so characteristic of inherited retinal diseases, are true for other inherited conditions remains to be seen. (Inherited deafness may be as complicated; Petit et al 2001.) As distressing as the complexity may be for clinicians and patients, though, the fact that we know so much about the genes and mutations causing retinal diseases is a testament to the exceptional progress made in recent years. Nonetheless, there is a long row to hoe before we have a full appreciation of the molecular causes of inherited retinal diseases, especially among non-Western populations, and this information is only the first step in understanding pathogenic mechanisms and in providing treatments and cures.

Progress in identifying RD genes

There are many ways to measure progress in this field, for example, by the number of genes cloned, the number of mutations identified, or the number of investigators involved. Perhaps the most meaningful measure of progress is implied by one of the goals articulated in the Five Year Plan of the Foundation Fighting Blindness in 2000, '... to foster research leading to identification of the underlying genetic cause in 95% of patients with inherited retinal dystrophy...' (FFB Planning Document 2000).

That is, the question is not simply how many genes and mutations have been found but, rather, in what fraction of patients can a cause be identified? More explicitly, the issue is not 'can mutations in this gene cause retinal disease' but, instead, 'do mutations in this gene cause disease in actual patients and, if so, in what fraction of patients'. This emphasis places the focus on prevalence, which is directly relevant to patient care, to the market for gene-specific treatments, and to the burden of these disorders on society.

In practice, disease-causing genes are usually identified without consideration of prevalence, say, by linkage mapping and cloning, and later screened in a large population of patients with appropriate phenotypes to establish prevalence. Therefore, one measure of progress in identifying RD genes is simply a count of known genes. The RetNet database maintains a list of cloned and/or mapped RD genes with information on clinical associations, the protein product, if known, and primary references and links to other sites *(http://www.sph.uth.tmc.edu/RetNet)*. Figure 1 is a graph of the progress to date in mapping and cloning RD genes and Table 1 is a summary of the clinical categories for these genes, both derived from RetNet.

Figure 1 is based on the date of publication of the first article that reported chromosomal mapping of a RD locus and the first article that reported identification of the underlying disease-causing gene. In the case of RD genes identified by candidate gene screening, the 'mapping' and 'cloning' dates are the date of the first publication reporting identification of disease-causing mutations. Table 1 is simply a 'head count' of known RD genes. Please note that the clinical categories can be misleading because different mutations in the same gene may cause different diseases, yet the Table counts each gene once only. In cases where a gene falls into more than one category, it is counted in the category in which it was first described. For a list of all genes in all categories, see RetNet *(http://www.sph.uth.tmc.edu/RetNet/sum-dis.htm#B-diseases)*.

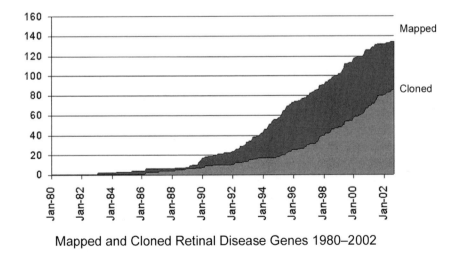

Mapped and Cloned Retinal Disease Genes 1980–2002

FIG. 1.   Mapped and cloned retinal diseases 1980–2002.

TABLE 1   Summary of selected retinal disease genes

Category	Mapped only	Cloned	Total
Total	46	87	133
Autosomal dominant retinitis pigmentosa	1	11	12
Autosomal recessive retinitis pigmentosa	5	10	15
X-linked retinitis pigmentosa	3	2	5
Usher syndrome	5	6	11
Bardet–Biedl syndrome	2	4	6
Leber congenital amaurosis	2	4	6
Autosomal dominant macular degeneration	4	4	8

Identification of RD genes progressed slowly from mapping the first X-linked retinitis pigmentosa (RP) gene in 1984 (Bhattacharya et al 1984) until the early 1990s, as seen in Fig. 1. Then, a number of technical developments, such linkage mapping using tandem repeat polymorphisms and improved software, greatly accelerated gene mapping. Within a few years thereafter the approaching completion of the human genome project provided tools to accelerate gene cloning. For most of the past 10 years, the number of mapped RD genes has increased roughly linearly and the number of cloned genes has increased more rapidly. From Fig. 1 there appears to be a flattening of both curves, around the time of publication of the first draft of the human genome (by coincidence or otherwise). As noted below, since there are clearly many RD genes yet to be identified, the work is not yet done. A reasonable prediction is that laboratory techniques for mapping and identifying RD genes will improve markedly in the near future. However, the rate limiting steps are in ascertaining patients and families, and in clinical evaluation. Therefore, as always, the clinicians who work with affected individuals are the critical ingredient for continuing progress.

To date, 133 RD genes have been mapped and 87 have been cloned, as seen in Table 1. The Table counts genes causing many different retinal diseases, but even within a limited category such as retinitis pigmentosa (RP) there are numerous RD genes. For example, mutations in 12 genes cause autosomal dominant RP (ADRP), mutations in 15 others cause autosomal recessive RP (ARRP), and mutations in 5 cause X-linked RP (XLRP). In total, more than 49 genes are associated with RP, if syndromic forms are included. This illustrates the exceptional genetic heterogeneity of inherited retinal diseases.

Even these numbers are an underestimate, because they do not account for gene mutations implicated in more than one category of disease. For example, mutations in peripherin 2 (*RDS*) may cause either ADRP or autosomal dominant macular degeneration (Felbor et al 1997), and mutations in usherin (*USH2A*) may cause either Usher syndrome or non-syndromic ARRP (Rivolta et al 2000).

Functional categories of RD genes

The proteins produced by the known RD genes fall into several functional categories, such as phototransduction, the visual cycle or retinal transcription factors (Phelan & Bok 2000, Rattner et al 1999). As might be expected, many RD genes are expressed exclusively or principally in photoreceptors and/or RPE cells, and play a role in 'obvious' visual functions. The limited number of relevant pathways, in comparison to the large number of RD genes, raises the hope that therapies may be focused on downstream components of these pathways rather than each individual RD gene. Therapies directed at slowing retinal cell apoptosis are an example of pathway targeting (Chader 2002).

However, several recently-identified RD genes are ubiquitously expressed, and do not fall comfortably into 'vision pathways'. Perhaps the most striking counter examples are three genes, HPRP3, PRPF8 (RP13) and PRPF31 (RP11). Mutations in these genes cause ADRP, and *only* ADRP (as far as is known), but each codes for a distinct, highly-conserved, essential protein component of the RNA splicing complexes found in all eukaryotes (Chakarova et al 2002, McKie et al 2001, Vithana et al 2001). If nothing else, these findings demonstrate that our understanding of retinal development, cytology and biochemistry is more limited than generally recognized. They also illustrate the power of RD gene identification to reveal new aspects of retinal biology.

Progress in identifying disease-causing mutations in RD genes

Table 2 presents the number of unique disease-causing mutations reported for several selected RD genes. The data are based on pathogenic mutations listed in the Human Gene Mutation Database (*http://archive.uwcm.ac.uk/uwcm/mg/hgmd0.html*, Krawczak et al 2000). Additional resources for RD gene mutations are the Mutation Database of Retina International (*http://www.retina-international.org/sci-news/mutation.htm*) and individual RD gene entries in OMIM (*http://www3.ncbi.nlm.nih.gov/Omim/*, Hamosh et al 2002).

To date, rhodopsin mutations are the most numerous, with more than 100 reported, but other RD genes, such as RDS and MYO7A, have more than 50 reported mutations each. There is a correlation between the time since the first mutation was reported and the current total for each gene—suggesting that the number of distinct mutations observed is related to the number of patients surveyed. Because of the large number of known mutations in some cases, there

TABLE 2   Summary of selected retinal disease mutations

Category	Gene symbol	Mutations reported
Autosomal dominant retinitis pigmentosa	RHO	103
	RDS	69
	RP1	14
	IMPDH1	9
Autosomal recessive Leber congenital amaurosis	RPE65	39
	GUCY2D	32
	AIPL1	14
X-linked retinitis pigmentosa	RPGR	62
	RP2	30
Usher syndrome	MYO7A	56
	USH2A	31
	USH1C	7

have been several attempts to find simplifying rules for the association between genotype and phenotype, with limited success so far. Some pathogenic mutations seem to cluster within protein domains (e.g. RP1; Berson et al 2001), whereas mutations in other genes are scattered throughout (e.g. rhodopsin; Hargrave 2001). In some cases loss-of-function mutations predominate, whereas, in others, missense mutations are more frequent (e.g. RP1 and rhodopsin, respectively). That is, global rules to explain the occurrence, molecular distribution and clinical consequences of RD gene mutations have not yet emerged.

The striking allelic heterogeneity of RD genes has disquieting implications. As noted below, some mutations in RD genes are 'common', relative to other mutations at the same locus, but the aggregate frequency of the rare or one-of-a-kind mutations is at least 50% in most cases. Thus to find all pathogenic mutations within a given gene it is necessary to scan the entire coding sequence, at least. This is a daunting task for large genes such as RP1 with 2156 amino acids or ABCA4 with 50 coding exons.

Of more concern, though, is that rare pathogenic mutations occur on a genetic background of numerous non-pathogenic variants, both polymorphic and rare. Simply because an amino acid substitution is found in a RD gene, and is rare, does not mean it is pathogenic. Worse, in positional candidate cloning of a mapped, dominant disease locus, simply because a rare missense mutation is tracking with disease does not prove pathogenicity. This is because rare, benign variants will be 'common', in aggregate, across the large non-recombinant regions usually implicated in positional cloning projects.

The extent of background variation is documented in a recent publication by Stephens et al (2001). The paper reports the numbers and types of sequence variants detected in sequencing 313 human genes in 82 individuals, representing 4 ethnic groups. A total of approximately 3900 polymorphic nucleotide substitutions (SNPs) were observed; 23% were variable within all groups, 25% were variable within one group only, and 38% were rare, i.e. found in one individual only. Of these, more than 50% lead to an amino acid substitution and roughly 1% introduce a premature stop codon. Put another way, in this study, on average

- an individual is heterozygous every 1.4 kb within coding sequences
- roughly half of these entail an amino acid substitution, and
- roughly 1 in 25 are rare (unique to an individual or family)

Thus rare, non-pathogenic amino acid substitutions will be found, inevitably, in screening projects in which the same gene is sequenced in many individuals, or in which several contiguous genes are sequenced in one individual. The conclusion: it is not sufficient to say that a mutation is pathogenic because it is rare — additional genetic and/or biochemical evidence must be adduced to make the case.

Prevalence of disease-causing mutations causing autosomal dominant retinitis pigmentosa

A misreading of Table 2 might suggest that all pathogenic mutations in RD genes are unique, one-of-a kind events. This is not usually the case. In fact, of the scores of mutations reported at each locus, typically 2 or 3 are much more prevalent than others. Focusing specifically on ADRP, Table 3 lists pathogenic mutations which are found in multiple, unrelated families (Sohocki et al 2001, Bowne et al 1999, 2002). (Similar findings apply to other forms of inherited retinal disease.) 'Unrelated' in this context means that families sharing a mutation are usually not aware of each other's existence but that, nonetheless, the mutation arose in a common ancestor, perhaps hundreds of years ago. That is, where haplotype data are available, most prevalent mutations have been shown to arise by founder effects, not recurrent mutation.

For each gene in Table 3, a few mutations account for a large fraction of the total. For rhodopsin, the Pro23His mutation accounts for 40% of the total; for RP1 and IMPDH1, one mutation in each case, Arg677ter and Asp226Asn, respectively, accounts for 50%. There is an important caveat to these observations: the prevalent mutations are always limited to a specific geographic group, consistent

TABLE 3   Common retinal disease mutations

Gene symbol	Mutation	% of total per gene
RHO	Pro23His	40
	Arg135Trp	3
	~5 others	6
RDS	Pro210Arg	25
	IVS2 A > T	12
RP1	Arg677ter	50
	Leu726del5	16
	Gly723ter	16
IMPDH1 (RP10)	Asp226Asn	50
	Gly324Asp	25

with historically recent founder events. For example, the rhodopsin Pro23His mutation is found in Americans of European origin but not Europeans (Farrar et al 1990). Thus any contributions to prevalence are limited to specific populations.

Progress in identifying mutations in patients with autosomal dominant retinitis pigmentosa

Taking adRP as a representative example of progress in identifying RD genes, it is now possible to identify a disease-causing mutation in 50–60% of patients. Table 4 shows the derivation of this estimate [references in RetNet]. As each ADRP gene was identified, the first reports were followed by screening — either by sequencing or mutation scanning (e.g. by SSCP or DGGE) of large collections of patients. Although the published studies are dissimilar in methodology, roughly comparable prevalences have been reported. Thus rhodopsin mutations are found in approximately 30% of adRP cases; mutations in four genes, PRPF31, RP1, RDS and IMPDH1, account for 5–8% each; and mutations in a few others account for 1–3% each. Table 4 lists the known ADRP genes in order of prevalence, with a commutative prevalence of 60%.

This represents great progress toward the goal of identifying the cause in 95% of affected individuals, but the estimates come with *many* caveats. First, the selection of patients and screening methods differ between studies. Perhaps the greatest difference is in defining 'autosomal dominant' families. Second, each study is relatively small, and little or no attempt has been made to estimate confidence intervals. Third, all such studies have subtle biases in ascertainment of patients which may limit applicability to unselected populations.

TABLE 4 Prevalence of identifiable mutations in patients with autosomal dominant retinitis pigmentosa

Gene symbol	% of ADRP	Commutative %
RHO	30	30
PRPF31 (RP11)	8	38
RP1	6	44
RDS	6	50
IMPDH1 (RP10)	5	55
HPRP3 (RP18)	3	58
NRL	1	59
others	< 1	60

Finally, the most important caveat is that, at best, the estimates are only applicable to the two major groups commonly studied: Americans of European origin and Europeans. Recent studies suggest that frequencies in Asia and elsewhere will be very different (Zhang et al 2002). This is explained, in part, by the predominance of founder mutations which are limited in geographic distribution. A reasonable prediction is that the genes and mutations identified in Americans and Europeans will be different from the most common causes in other populations. Further, of course, additional major genes will be identified in the 'well-studied' groups, too.

Implications and conclusions

By any reasonable measure, substantial progress has been made in identifying genes and mutations causing inherited retinal diseases. In spite of the exceptional heterogeneity observed, it is possible to detect disease-causing mutations in upwards of 60% of cases in certain well-defined patient populations, such as Americans with autosomal dominant retinitis pigmentosa.

Unfortunately, there is a very large gap between what can be done in theory and what is possible in practice. Detection of mutations in 60% of ADRP patients requires sequencing, or the equivalent, of six or more RD genes, totalling more than 60 000 kb of DNA over 40 exons. Any reasonable estimate of costs is in the thousands of dollars per patient (Table 5). Pre-screening for the most common mutations is much less expensive but will miss, perhaps, 50% of pathogenic variants. At present, routine mutation testing is not available for ADRP patients, either commercially or in an academic setting. Developments in technology will eliminate this diagnostic bottleneck within a few years, we hope.

TABLE 5 Estimate of costs for mutation screening of patients with autosomal dominant retinitis pigmentosa

Item	Cost ($US)	Commutative costs
Set patient file	100	100
Prepare and store DNAs	200	300
Screen for 'common' mutations by DHPLC	500	800
Sequence genes causing > 5% of ADRP	2000	2800
DHPLC and sequence remaining genes causing > 1%	3500	5300

The era of gene mapping and positional 'cloning' is not over, notwithstanding the great progress to date. Whether only a few new RD genes will bring the total to 95% of patients, or many, is unknown. It is very likely that different sets of RD genes will predominate in other populations. That is, we have achieved much but we have a long way yet to go.

References

Berson EL, Grimsby JL, Adams SM et al 2001 Clinical features and mutations in patients with dominant retinitis pigmentosa-1 (RP1). Invest Ophthalmol Vis Sci 42:2217–2224

Bhattacharya SS, Wright AF, Clayton JF et al 1984. Close genetic linkage between X-linked retinitis pigmentosa and a restriction fragment length polymorphism identified by recombinant DNA probe L1.28. Nature 309:253–255

Bowne SJ, Daiger SP, Hims MW et al 1999 Mutations in the RP1 gene causing autosomal dominant retinitis pigmentosa. Hum Mol Genet 8:2121–2128

Bowne SJ, Sullivan LS, Blanton SH et al 2002 Mutations in the inosine monophosphate dehydrogenase 1 gene (MPDH1) cause the RP10 form of autosomal dominant retinitis pigmentosa. Hum Mol Genet 11:559–568

Chader GJ 2002 Animal models in research on retinal degenerations: past progress and future hope. Vision Res 42:393–399

Chakarova CF, Hims MM, Bolz H et al 2002 Mutations in HPRP3, a third member of pre-mRNA splicing factor genes, implicated in autosomal dominant retinitis pigmentosa. Hum Mol Genet 11:87–92

Farrar GJ, Kenna P, Redmond R et al 1990 Autosomal dominant retinitis pigmentosa: absence of the rhodopsin proline–histidine substitution (codon 23) in pedigrees from Europe. Am J Hum Genet 47:941–945

Felbor U, Schilling H, Weber BHF 1997 Adult vitelliform macular dystrophy is frequently associated with mutations in the peripherin/RDS gene. Hum Mutat 10:301–309

FFB Planning Document 2000, Foundation Fighting Blindness, Owings Mills, MD, USA

Hamosh A, Scott AF, Amberger J, Bocchini C, Valle D, McKusick VA 2002 Online Mendelian Inheritance in Man (OMIM), a knowledgebase of human genes and genetic disorders. Nucleic Acids Res 30:52–55

Hargrave PA 2001 Rhodopsin structure, function, and topography: the Friedenwald lecture. Invest Ophthalmol Vis Sci 42:3–9

Krawczak M, Ball EV, Fenton I et al 2000 Human gene mutation database — a biomedical information and research resource. Hum Mutat 15:45–51

McKie AB, McHale JC, Keen TJ et al 2001 Mutations in the pre-mRNA splicing factor gene PRPC8 in autosomal dominant retinitis pigmentosa (RP13). Hum Mol Genet 10:1555–1562

Petit C, Levilliers J, Hardelin JP 2001 Molecular genetics of hearing loss. Annu Rev Genet 35:589–646

Phelan JK, Bok D 2000 A brief review of retinitis pigmentosa and the identified retinitis pigmentosa genes. Mol Vis 6:116–124

Rattner A, Sun H, Nathans J 1999 Molecular genetics of human retinal disease. Annu Rev Genet 33:89–131

Rivolta C, Sweklo EA, Berson EL, Dryja TP 2000 Missense mutation in the USH2A gene: association with recessive retinitis pigmentosa without hearing loss. Am J Hum Genet 66:1975–1978

Sohocki MM, Daiger SP, Bowne SJ et al 2001 Prevalence of mutations causing retinitis pigmentosa and other inherited retinopathies. Hum Mutat 17:42–51

Stephens JC, Schneider JA, Tanguay DA et al 2001 Haplotype variation and linkage disequilibrium in 313 human genes. Science 293:489–493

Vithana EN, Abu-Safieh L, Allen MJ et al 2001 A human homolog of yeast pre-mRNA splicing gene, PRP31, underlies autosomal dominant retinitis pigmentosa on chromosome 19q13.4 (RP11). Mol Cell 8:375–381

Zhang X, Fu W, Pang CP, Yeung KY 2002 Screening for point mutations in rhodopsin gene among one hundred Chinese patients with retinitis pigmentosa. Zhonghua Yi Xue Yi Chuan Xue Za Zhi 19:463–466

DISCUSSION

Kaleko: What percentage of RP mutations are spontaneous, without family history? And are you trying to find a moving target of new mutations, for which you cannot rely on family histories, that may prevent you from reaching your 95% goal?

Daiger: The answer to your second question is that we won't know until we approach closer to that figure. There is a formal possibility that we will be dealing with increasingly rare events that will be hard to find. For all of these diseases that fall into the 'common event' category, almost by definition those are not new mutations. They are mutations that have accumulated in a population. Frankly, we wouldn't see them if there weren't multiple families. Multiple families means by definition that the disease is passed from generation to generation. In our anecdotal experience we have never seen a genuinely new mutation. Take this with a grain of salt though, because we have seen nucleotide changes that don't appear in the parents. We also have a couple of cases of *de novo* amino acid substitution which we don't think are the cause of the disease. The bottom line is that none of us are looking for or anticipating new mutations.

Bolz: It is also interesting to look at other genes that influence the disease process and not just the disease genes. For example, it has been shown in autosomal dominant deafness and RP that family members can have different phenotypes or that they don't show the disease at all. In RP11 families, some people carry the mutation but don't show any phenotype (Vithana et al 2001).

Daiger: I agree. I suspect that the next wave in this field will be looking at modifying genes. We are particularly interested in the allele in *trans* to a disease-causing dominant mutation. For example, a mutation of the *RP1* gene is very common in a large family in Kentucky. The 'wild-type' chromosome has a tremendous amount of haplotypic variation at the amino acid level. We think there are at least six different haplotypically distinct proteins in that family alone. The obvious thing to look at for modifying factors, at first, are the alleles in *trans*, and then to look at the protein haplotypes that are occurring in other loci. I didn't mention the digenic inheritance for *ROM1* and *RDS*, which is a complication. It almost scares me in some ways: I don't like to address the fact that we have only scratched the surface of the degree of complexity of the diseases we are dealing with, at least at a genetic level.

Farber: Alan Bird, could you say something about the correlation between phenotype and genotype?

Bird: A lot of the disorders are quite distinct. RDS172 can be recognized clinically quite clearly. In maternally inherited diabetes the chromosomal disorder is absolutely diagnostic clinically. When you use the term 'RP', you are talking about generalized retinal dystrophies. Fewer than 50% of people labelled as having RP have RP as described in a textbook.

Daiger: The underlying clinical heterogeneity makes these categories very amorphous.

Bird: At least 50% of those autosomal dominant RP patients will lose central vision long before they have small visual fields. They have a physically separate macular atrophy. The term 'RP' is fine, as long as people don't consider it exclusively to be a disorder of rods in which central vision remains good until the visual field is minuscule.

Daiger: This is a fundamental philosophical problem. We have fallen into this trap of believing that the words we use contain biological meaning. We would probably be better off thinking about each of these diseases as an *n*-dimensional space for each of the clinical features: visual fields and so on. At one end there is RP, at the other there is Leber congenital amaurosis (LCA), and in the middle there is a gradation between the two.

Bird: In LCA, although people consider this as blindness from birth this is not the case. In Leber's original paper he said that this was early-onset recessive RP. He didn't call it congenital blindness. There may be a small number of people blind from birth, but it is a very small proportion of the total with this diagnosis.

Daiger: As a minimum, the disease should be defined as a particular mutation in a particular protein, but even this is not sufficient because of the tremendous variation within families. The disease is what this individual has, considered against the history of what we know of this gene in other members of the family.

Bird: When we do accurate phenotyping we can often guess the gene involved.

Daiger: I agree, in the context where some mutations are more common than others in the population that is ascertained. This wouldn't work if you went off to Gambia and tried to do the same with an African population, but it works quite well in London, for example.

Zack: Alan Bird's comments raise an issue that has come up before. If we are going to move ahead in our understanding of the effects of modifier genes, we need a diagnostic set of genotype/phenotype correlation tests. It would be helpful if we could devise tests for every patient in a genetic study, plus have a database on the Internet containing the actual values for the ERGs, for light adaptation, or whatever measures the clinicians feel are most important. If these data are available to everyone who does genetic studies then these correlations could happen.

Dryja: With dominant RP, the responsible genes for about 60% of cases of RP in the USA are identifiable. But to actually find which is the responsible gene in a particular patient requires the exon-by-exon SSCP or DNA sequencing method. Nobody augments this method with Southern blotting. If you look at the retinoblastoma gene for comparison, all retinoblastoma cases are known to be caused by the same gene. If you do exon-by-exon sequencing and Southern blotting, you find only about 70% of the mutations, and the remaining mutations are in introns or promoter regions.

Daiger: In my lab we look very carefully for homozygosity across a long DNA stretch, because this is often suggestive of a deletion that is actually hemizygous. This doesn't bring up regulatory mutations: in practice, no one knows how to look for these at a sequencing level. So there is the formal possibility that a fraction will be picked up. This is one of the reasons that we are interested in continuing allelic studies even on the known disease causing genes. Just because we fail to find a mutation in rhodopsin doesn't mean that there isn't something that we simply haven't been smart enough to see.

Dryja: My other point refers to your economic analysis of diagnostic testing. Perhaps the real success for diagnostic laboratories for retinal degenerations will occur once there is a gene-specific therapy. All families with RP will then want to know whether their disease is due to the gene for which there is a therapy, and this will create an explosion in the demand for DNA diagnostic laboratories. Until then, the cost–benefit is very high when one considers that it takes $5000 to do the test and there is very little benefit to knowing one's mutation.

Daiger: The cost per gene, if it is a small gene such as rhodopsin, is relatively modest. If RPE65 is the mutation where gene therapy really starts to work in humans, then I'm sure that commercial laboratories will take on RPE65 screening.

Bird: Your costing was so much per new dominant family. If you have a large pedigree in which the mutation is known, analysis is cheap. In a way that figure is the cost per family in dominant disease. If you have really good genetic registers it then becomes quite cheap.

Daiger: Note that included in this figure is the cost to establish a file on a patient. That is, just to set up a file on the patient with the legal paperwork costs $100. The overhead costs and the cost of making DNA and storing it is another $200. Thus we are looking at a cost of $300 before we detect a mutation. Nonetheless, if you tell me that a particular family has a codon 23 mutation in rhodopsin, then for another $50 plus the $300 set-up fee they can be added.

Swaroop: You talked about founder effects and mentioned that 40% of rhodopsin mutations in the USA can be accounted by P23H. How much of this is based on the haplotype analysis?

Daiger: Almost all of it, to the extent that you can do haplotype analysis. For rhodopsin, the codon 23 mutation is unequivocally a founder effect. Where it is not is the 135 mutation: there are at least two or three different haplotypes associated with this mutation. It is as if there is a target for mutation and by our ascertainment methods we are picking up more than you would expect simply by mutation alone.

Swaroop: Is this true for Arg677 and Asp226 mutations as well?

Daiger: I don't want to get trapped into saying specifically whether this is the case for each one. I know it is absolutely true for the *RP1* mutations. We have crude evidence that it is true for *IMPDH1*, but it isn't strictly true across the board for those five others. There are in fact recurring mutations within that category.

McInnes: From examining this do you have the impression that any of the so-called 'mutant' alleles out there are not well documented as pathogenic? I realise that in the papers in which they were originally established the evidence is usually pretty good, based on inheritance.

Daiger: By definition we are looking at mutations that have a penetrance of 98–99%, and an expressivity that is clear by the age of 40. It has to be true that there are other alleles in the population with lower penetrance or more variable expressivity that we simply can't find. Whether we call them 'modifiers' or the accumulation of old age mutations is another issue. One of the hopes is that age-related macular degenerations will turn out to be a consequence of modestly common alleles, the total effect of which is either additive or interacts in some other way. The huge technical problem of any kind of statistical analysis is the massive background variation in human populations. We all differ in thousands of amino acid substitutions. It is very hard to wade through this, and this is why we concentrate on very high penetrance alleles with fairly narrow expressivity.

Travis: What about situations where a mutation maps to a particular locus, yet we never find a mutation there? This seems to vary from gene to gene. Should we say that the gene in question is affected and that we just can't find the mutation, or should we consider the possibility of a linked locus?

Daiger: Anand Swaroop is probably the best person to answer this. At least for a fraction of those for which we fail to find mutations, eventually once we have all the splice variants and we understand the full-length cDNA, some of those will turn out to have mutations. Part of it is that we don't understand the structure of genes well enough yet.

Cremers: We looked at the choroideraemia gene and found most of the mutations in that gene. If one performs an extensive mutation analysis, including RT-PCR, one can find $\sim 10\%$ of the mutations in the introns that yield cryptic exons that are spliced into the mRNA.

Travis: In the case of choroideraemia there is no evidence that another locus is involved. In this case, for what percentage of choroideraemias can you find a molecular defect?

Cremers: For those patients we have tested rigorously, 95%. It is easier with choroideraemia because one-quarter of patients carries a deletion, and all mutations are stop mutations.

Dryja: Choroideraemia is hemizygous, so the deletions are easily detectable even by the exon-by-exon technique.

Swaroop: I would add that there are lots of variations that are seen in the untranslated regions or the intronic regions, so how can you say these are real mutations unless you do some kind of functional assay? Even though we do find sequence changes we cannot unequivocally say that these are causative gene mutations. This is a major problem. There are also difficulties with unidentified exons. *RPGR* is a classic example of this: this gene has multiple alternately spliced transcripts and a number of new exons have been discovered after the original gene cloning. A further problem concerns clinical diagnosis. Unless you have done genetic linkage in a family, it is very hard to say that a region is clearly a particular disease locus. With regard to RP, *RPGR* accounts for at least 70% of mutations. I am very conservative in this respect; others might say 90%.

McInnes: There were several papers presented at a recent meeting of the American Society of Human Genetics that reminded us that 60% of sequences conserved between mouse and human are not genic (i.e. not within the transcription unit *per se*). There is a lot of conserved, and therefore important, sequence that has to do with regulation of expression. Also, an increasing number of mutations are incredibly distant. For example, there are some *Shh* mutations that are single base pair changes in regulatory elements that are a megabase away from the gene, and they are embedded in another gene. A lot of these things are going to be very hard to find. We will need functional assays.

Swaroop: You can do such experiments in *Drosophila* and mouse to prove that these changes are really causative, but not in humans.

Farber: Eventually, I think we will have to develop fast methods to express the mutations and see whether they are really causing the disease. Sometimes we find some variations in a gene that seem to be the ones responsible, but when tested they are not.

Swaroop: One more comment. You talked about conservative versus non-conservative changes. In another gene *NRL*, for example, the first change we found was serine to threonine. It was hard for us to convince ourselves (and others) that this was a real disease-causing mutation. Shomi Bhattacharya and I struggled with the data until we had functional assays, when we were able to show that this conservative substitution can affect the NRL's transactivation function. Hence, there could be small 'conservative' changes that are disease causing; on the other hand, there could be what appear to be very dramatic changes that don't result in the disease. You also mention about these 'stop codons'. This is shocking, and if this is indeed the case we are in deep trouble. Unless we have a functional assay we cannot say that a particular mutation is disease causing.

Daiger: This is why *Nature Genetics* has stopped taking papers that don't have functional proof that a mutation causes disease.

Thompson: I would like to make a case for the importance of continuing to support and focus our efforts on basic studies of biochemistry and cell biology as a means to identify both rare mutations and modifier genes involved in retinal degeneration. If the situation is such that many of the mutations are present in isolated populations as rare defects, we will not be able to find all the disease genes easily. Knowing the biochemistry will be the fastest track to getting us to a functional assay that we can use in the laboratory, and then getting to a therapy from there. The genetic studies are fabulously important, but continued studies of the biology and chemistry of the photoreceptors will also be of vital importance

Chader: I agree, and at The Foundation Fighting Blindness this is a core emphasis. We call this approach 'back to the future': the biochemistry and cell biology are important not only for finding new genes, but also once the gene is found this is just the first step towards developing a therapy. The cell biology and all the sequelae after that are very important, as is putting together a compendium of the information known prior to the gene finding and after the gene finding. The way we try to approach this is to try to put together small consortia of investigators. We are hoping to establish more of a cooperative venture around the various disease gene entities, which we think will enhance the whole process.

Bhattacharya: Steve Daiger, you mentioned linkage disequilibrium mapping and regions of linkage disequilibrium in the human genome, and whether this will be of use in relation to complex disease such as age-related macular degeneration (AMD). Do others have comments about how we could identify genetic loci using this technique?

Daiger: Among people who are looking at genes that increase the risk of common diseases such as type 2 diabetes, the underlying paradigm is that there will be alleles in the population that are modestly common, at 1–5%, hence by the old definition 'polymorphic'. The idea is that these alleles will increase or decrease the risk. If those alleles arose on some sort of chromosomal background then when you ascertain everybody who has type 2 diabetes with that background, which is essentially the extended haplotype, it will lead you to that mutation. This is why all the single nucleotide polymorphism (SNP) analysis is being done. If it turns out that the increased risk for AMD is a few missense changes or other kinds of changes, then linkage disequilibrium will find it. But then there is a conceptual problem: the one gene that has been put forward as contributing to AMD (with controversy) is the *ABCA4* gene (which used to be called *ABCR*). What this group did was to look at AMD patients, and they found a huge amount of background, genuinely polymorphic, variation. They disregarded this. At the same time, they found a lot of rare variants which they believed were significantly increased in frequency in affected individuals versus non-affected individuals. If you look at this from a population genetics point of view, there is a huge red flag, which is that a gene that tolerates background polymorphic variation will also tolerate an increased number of these rare variants. This case has ended up in limbo. With credit to everyone involved, the statistical measures simply don't exist to analyse this. However, the warning here is that if AMD is the result of the accumulation of very rare mutations, then linkage disequilibrium will not find it, nor will linkage analysis. All we can do is to do candidate gene screening. This has huge, inherent statistical problems. This is a much more difficult field than it appears to be superficially. I am convinced, though, that if you went into Americans of Western European origin and looked at autosomal dominant RP, you could detect the rhodopsin codon 23 mutation by linkage disequilibrium over a range of about a million base pairs.

Bhattacharya: At the moment a linkage disequilibrium (LD) map of the human genome is under construction. The caveat is that it will be important also to think about generating LD maps in different populations and not just caucasians.

Daiger: The wild card in all this is the discovery that the LD blocks in humans are on the order of about 50 000–100 000 base pairs. This represents our evolutionary history perhaps over the last 20 000–50 000 years. It is probably true that if we identify a subset of SNPs within each one of those 50 000 base pairs using microarray technology we could quickly scan 1000–2000 individuals. I see this technology coming online within a couple of years, and I would like to see it applied to some of the genes modifying inherited retinal diseases.

Bhattacharya: That is an important application and it will focus our work in relation to mapping complex traits.

Daiger: The problem is that it is currently expensive.

Kaleko: Since it will be very difficult to introduce numerous, gene-specific gene therapies, it will be necessary to find therapeutic strategies that will treat multiple diseases. You have been able to categorize your mutations into common pathways, such as visual cycle and transcription factors. Do any of those common pathways suggest alternative treatment strategies that might be independent of the particular mutations?

Daiger: I'm the wrong person to answer that. Certainly, the conventional wisdom is that all of these pathways eventually lead to apoptosis. Recent evidence that there may be two different apoptotic pathways makes me a little nervous about this.

Bok: I can give an example of a therapy independent of the particular mutation, which is the RPE65 knockout mouse. This has a missing protein that seems to be essential for the production of the chromophore used in vision, which is 11-*cis*-retinaldehyde. Palczewski and collaborators have provided the chromophore in the diet and restored some function in these animals. The problem with this nutritional approach is that it is a somewhat toxic molecule.

Kaleko: I was actually referring to gene therapy. Are there second site genes that could be used to treat many different types of mutations?

Daiger: For example, several inherited retinal diseases have high cyclic nucleotide levels as an intermediate step in the disease process. Could you just lower guanine levels and deal with several diseases like that?

Bok: For RPE65, if the mutation leads to a sluggish or non-functional RPE65, irrespective of where the mutation is, you can do replacement gene therapy readily because these mutations are recessive.

Dryja: What Michael Kaleko is asking is, if we had a cure for RP65, could we envision that this method of cure might help the retinal degeneration caused by mutations in genes encoding other components in the same pathway, such as *LRAT* or *ABCR*?

Swaroop: Theoretically, it is possible that many of these mutations work through similar pathways to lead to retinal degeneration. The end point in all these cases is the death of the photoreceptors. We have hypothesized that many of the pathways will converge at some point before they lead to the apoptotic pathways. If this is the case, we should be able to target the molecules that are in between.

Bok: That is a different question. Many of us are using this approach: we want to know what the genes are that then respond to that mutation and lead you to apoptosis.

Aguirre: The question deals with the pathways, and these are functional pathways and not pathways for disease. We may need to develop a different perspective. We need to examine these genes and see how they are acting in a common disease pathway.

Daiger: I would second what Anand Swaroop said. There is still the hypothetical hope that several of the phototransduction mutations will lead to a common biochemical abnormality, the end of which is apoptosis. We can deal with that common abnormality. It is more likely that we will find this with enzymatic systems as opposed to structural mutations.

Farber: The same thing happens with phototransduction. Depending on where in the phototransduction genes the mutation is — upstream or downstream — you may be able to do something.

Dryja: Phototransduction in rods is different from that in cones. You could treat all the forms of RP caused by mutations in phototransduction genes that are specific to rods, if you could find a therapy that saves the cones. In other words, in the forms of RP caused by mutations in rod-specific genes, the cones are dying secondarily since the mutant gene is not expressed by the cones. If someone discovers an agent that prevents the cones from dying, it should help all of the rod-specific forms of RP.

Molday: This is where apoptosis is critical. If we can stop the rods from dying by inhibiting the apoptotic pathway, even if they are not functional this may save the cones from dying as long as the genetic defect is specific for rods and not cones. An individual can lead a relatively normal life utilizing only cones as their only functional retinal photoreceptor cell.

Thompson: This was Paul Sieving's idea behind the idea to use Accutane for therapy.

Chader: The term 'gene-based pharmaceutical therapy' has been used. There are many steps that control Ca^{2+} levels, for example. If you can control the Ca^{2+} level through gene therapy or a drug, this may give you a magic bullet that obviates the individual problems in those genes.

Kaleko: As someone who has done a lot of gene therapy but not much in retinal disease, I find it daunting that there are so many mutations. There will have to be common pathways that we can treat.

Bok: This is why the ciliary neurotrophic factor (CNTF) approach is attractive. It ameliorates many of the mutations, or at least gives a palliative effect. I can also give an example of a problem I am working on where it seems that acid–base balance is a problem. Knocking out a particular bicarbonate transporter kills photoreceptors and affects hearing. I assume this is a pH problem in the cell. If this could be controlled it might be an effective therapy.

Reference

Vithana EN, Abu-Safieh L, Allen M J et al 2001 A human homolog of yeast pre-mRNA splicing gene, PRP31, underlies autosomal dominant retinitis pigmentosa on chromosome 19q13.4 (RP11). Mol Cell 8:375–381

Dominant cone and cone-rod dystrophies: functional analysis of mutations in retGC1 and GCAP1

David M. Hunt, Susan E. Wilkie, Richard Newbold*, Evelyne Deery*, Martin J. Warren*, Shomi S. Bhattacharya and Kang Zhang†

*Institute of Ophthalmology, University College London, Bath Street, London EC1V 9EL, UK, *School of Biological Sciences, Queen Mary, University of London, London E1 4NS, UK and †Department of Ophthalmology and Visual Science, and Program in Human Molecular Biology & Genetics, University of Utah, Salt Lake City, UT 84112, USA*

Abstract. The regulation of cGMP levels is central to the normal process of phototransduction in both cone and rod photoreceptor cells. Two of the proteins involved in this process are the enzyme, retinal guanylate cyclase (retGC), and its activating protein (GCAP) through which activity is regulated via changes in cellular Ca^{2+} levels. Dominant cone-rod dystrophies arising from changes in retGC1 are essentially restricted to mutations in codon 838 and result in the replacement of a conserved arginine residue with either cysteine, histidine or serine. In all three cases, the effect of the substitution on the *in vitro* cyclase activity is a loss of Ca^{2+} sensitivity arising from an increased stability of the coiled-coil domain of the protein dimer and retention of cyclase activity. In contrast, mutations in the Ca^{2+}-coordinating EF hands of GCAP1 result in dominant cone dystrophy; the consequences of these mutations is a reduced ability of the mutant protein to regulate retGC activity in response to changes in Ca^{2+} levels. Functionally therefore, the retGC1 and GCAP1 mutations are similar in reducing the feedback inhibition of Ca^{2+} on cyclase activity and thereby on cGMP levels in the photoreceptors.

2004 Retinal dystrophies: functional genomics to gene therapy. Wiley, Chichester (Novartis Foundation Symposium 255) p 37–50

The regulation of cGMP levels is central to the normal process of phototransduction in both cone and rod photoreceptors. After excitation by a photon of light, an enzymatic cascade of events occurs which leads to the activation of the enzyme cGMP-phosphodiesterase (PDE), the hydrolysis of cGMP and the closure of the cGMP-gated cation channels. This results in the hyperpolarization of the plasma membrane and the generation of a signal higher up in the visual pathway. On closure of the ion channel, the cytosolic levels of Ca^{2+}

decrease because export by the Na^+, K^+, Ca^{2+} exchanger continues. This reduced $[Ca^{2+}]_{free}$ results in the activation of retinal guanylate cyclase (retGC) by activating proteins (GCAPs). The increased conversion of GTP to cGMP leads to a restoration of cGMP levels to their dark level.

Changes in the regulation of cGMP are associated with mutations in all three components of this process. The *rd* mutation in the mouse is caused by a mutation in PDE (Pittler & Baehr 1991) and disease-causing mutations in both the α and β subunit genes have been reported in human (McLaughlin et al 1993, Dryja et al 1999). Mutations in GCAP result in dominant cone dystrophy (Payne et al 1998, Wilkie et al 2001) whereas mutations in retGC1 are responsible for recessive Leber congenital amaurosis (LCA1) (Perrault et al 1996) and dominant cone-rod dystrophy (Kelsell et al 1998).

The recovery process of photoreceptors after light exposure is mediated via a change in intracellular $[Ca^{2+}]_{free}$ in response to a drop in cyclic GMP level, both fundamental secondary messengers in photoreceptors. Several mechanisms, highly conserved across species, have developed to maintain a low $[Ca^{2+}]_{free}$ within photoreceptors. In the dark state, the $[Ca^{2+}]_{free}$ in photoreceptor outer segments rises to around 500 nM due to the influx of Ca^{2+} through open cGMP-gated Na^+/Ca^{2+} channels (Dizhoor & Hurley 1999). Exposure to light leads to the hydrolysis of cGMP by PDE (via the rhodopsin-mediated cascade), and a subsequent decrease in $[Ca^{2+}]_{free}$ to approximately 30 nM due to the continued efflux of Ca^{2+} through light-independent $Na^+/K^+/Ca^{2+}$ exchangers (Baylor 1996, Pugh 1996). This decrease in $[Ca^{2+}]_{free}$ stimulates an increase in cGMP, synthesized by retinal-specific guanylate cyclase, in a feedback mechanism vital to the process of light adaptation and the recovery of the dark state.

The retGCs are not directly sensitive to changes in $[Ca^{2+}]_{free}$; Ca^{2+} exerts its regulatory effect via specialized Ca^{2+} binding proteins known as guanylate cyclase activating proteins (GCAPs). Three mammalian isoforms have been identified, GCAP1 (Palczewski et al 1994), GCAP2 (Lowe et al 1995, Gorczyca et al 1995) and GCAP3 (Haeseleer et al 1999). GCAPs belong to a subgroup of the neuronal-specific Ca^{2+} binding proteins, which includes recoverin, all of which incorporate four variably functional repeats of the EF hand domain (Polans et al 1996), a helix-loop-helix structural domain with a selectively high affinity for Ca^{2+} binding. In the case of the GCAPs, only EF2, EF3 and EF4 are functional. Numerous studies have demonstrated that GCAPs function to mediate the Ca^{2+}-sensitive synthesis of cGMP by retGC, activating the cyclase at $[Ca^{2+}]_{free}$ < 100 nm and inhibiting the cyclase at $[Ca^{2+}]_{free}$ > 500 nm, concentrations characteristic of light- and dark-adapted photoreceptors respectively (Pugh et al 1997). This distinguishes the GCAPs from other Ca^{2+} binding proteins, which typically activate effector proteins in their Ca^{2+}-loaded forms.

The three GCAP isoforms have all been localized to the retina in various species, with GCAP1 present in higher concentrations in cone than in rod outer segments (Gorczyca et al 1995, Cuenca et al 1998), GCAP2 localizing to rods (Lowe et al 1995), cone inner segments (Otto-Bruc et al 1997) and layers of the inner retina (Cuenca et al 1998), and GCAP3 specific to cones (Imanishi et al 2002). To date two dominant cone dystrophy mutations have been identified in *GUCA1A*, the gene encoding GCAP1, a Tyr99Cys substitution in EF3 (Payne et al 1998) and a Glu155Gly substitution in EF4 (Wilkie et al 2001).

Mutations in *GUCY2D*, the gene encoding retGC1, have been shown to be responsible for LCA1 (Perrault et al 1996), the most severe form of inherited retinopathy with total blindness or greatly impaired vision recognized at birth or in early infancy. These mutations show a recessive pattern of inheritance with no reported heterozygous effects. In addition, mutations causing autosomal dominant cone-rod dystrophy have been identified in this gene. In contrast however to the LCA1 mutations that are found in most regions of the gene, the autosomal dominant cone-rod dystrophy mutations all share a common feature of the substitution of Arg838 (Payne et al 2001).

Experimental procedures

RetGC1 activity

Mutant and wild-type retGC1 activity was assayed as described by Wilkie et al (2000). Point mutations were introduced into a *retCC1* cDNA cloned in pBluescript using the Altered Site kit (Promega). The wild-type and mutant copies were then subcloned into the expression vector pRC-CMV (Invitrogen) and used for the transient transfection of HEK 293T cells. RetGC1 was isolated in the form of a membrane preparation. Measurement of cyclase activity followed the radioassay of Dizhoor et al (1995). Basal activity and activity stimulated by wild-type and mutant GCAP1 were measured in reaction buffers containing a range of $[Ca^{2+}]_{free}$.

Generation of recombinant wild-type and mutant GCAP1

A wild-type human GCAP1 cDNA clone used for the activity studies carried an artificial Glu6Ser substitution to facilitate N-myristoylation of the protein. The Glu155Gly and Tyr99Cys mutations were introduced with the GeneEditor site-directed mutagenesis kit (Promega). Mutant and wild-type GCAP1 was subcloned into pET3a (Novagen), expressed in *E. coli* strain BL21 and purified as described in Newbold et al (2001).

Results

Functional analysis of mutant GCAP1 activity

The binding of Ca^{2+} to the three EF hands of GCAP1 regulates its ability to activate retGC1. Thus at $[Ca^{2+}]_{free}$ below about 100 nM, retGC1 is activated by GCAP1 (Gorczyca et al 1994, Dizhoor et al 1994) but at micromolar concentrations it is inhibited (Rudnicka-Nawrot et al 1998). Ca^{2+} titrations of retGC1 activity in the presence of wild-type and Glu155Gly mutant GCAP1 are shown in Fig. 1A. With wild-type GCAP1, the $[Ca^{2+}]_{free}$ for half-maximal activity (ED_{50}) was estimated to be about 350 nM. With mutant GCAP1, however, the curve shows very little inhibition of activity above 350 nM and even at 1μM, almost 70% of maximal activity persists, compared to about 10% with wild-type. Using an equimolar mixture of wild-type and mutant GCAP1, activity at $[Ca^{2+}]_{free}$ above 350 nM is again maintained; at 1μM almost 50% of maximal activity persists. Thus the effect of the mutation is to maintain high levels of cyclase activity over the whole physiologically relevant range of $[Ca^{2+}]_{free}$. This effect is dominant in that it persists even in the presence of wild-type GCAP1 protein.

The Tyr99Cys mutation in GCAP1 has been shown to result in a similar constitutive activation of retGC1 at high $[Ca^{2+}]_{free}$ (Sokal et al 1998, Dizhoor et al 1998). Comparison of the magnitude of the effects of the two mutations on retGC activity (Fig. 1B) shows that the Glu155Gly mutation is more severe than Tyr99Cys in that the failure to inhibit activity at high $[Ca^{2+}]_{free}$ is more pronounced.

Functional analysis of mutant retGC1 activity

A series of point mutations at codon 838 were made in the retGC1 cDNA to generate the Arg838Ser/His/Cys mutations identified in the dominant cone-rod dystrophy patients (Kelsell et al 1998, Payne et al 2001). These were transiently transfected into HEK 293-T cells for expression and the activities of the recombinant proteins assayed *in vitro*. The three mutants all showed a somewhat depressed basal activity. However, when activated with recombinant 8 μM human GCAP1, the activities of the mutants were similar to wild-type (Fig. 2). The major difference in the activity of the mutant cyclases lies in their sensitivity to increasing $[Ca^{2+}]_{free}$. At $[Ca^{2+}]_{free} < 100$ nM, the mutants show a similar activity to wild-type but as the $[Ca^{2+}]_{free}$ increases, the cyclase activity remains elevated compared to wild-type and this persists beyond the normal physiological range of 100–500 nM (Fig. 3).

Domain structure

The retGC1 protein comprises five domains, an extracellular domain, a transmembrane domain, a kinase homology domain, a dimerization domain and

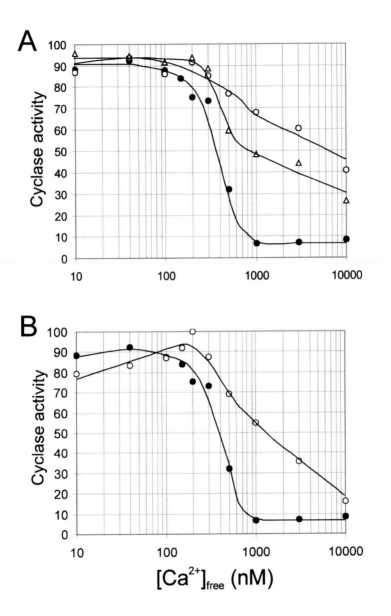

FIG. 1. Ca^{2+} sensitivity of activation of wild-type retGC1 by wild-type and mutant GCAP1. (A) Glu155Gly mutant GCAP1. (B) Tyr99Cys mutant GCAP1. Wild-type GCAP1, filled circles; mutant GCAP1, open circles; 50:50 mixture of wild-type and mutant GCAP1, triangles. Cyclase activity is expressed as percentage maximal activity. The assay contained 8 μM GCAP1 in all cases.

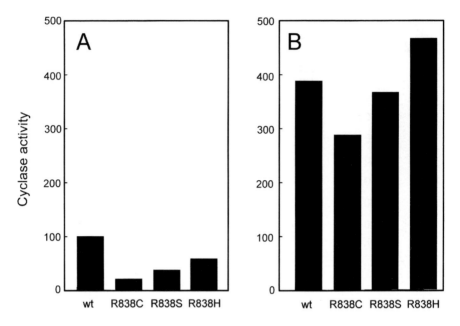

FIG. 2.   Basal and stimulated activities of wild-type and mutant retGC1. (A) Basal activity with zero free Ca^{2+}. (B) Activity with 8 μM GCAP1 with zero free Ca^{2+}. Cyclase activity is expressed as percentage basal activity of wild-type.

a cyclase domain. A significant feature of the distribution of mutations within the gene is that those causing recessive LCA1 are found in each domain except the dimerization domain, whereas the dominant cone-rod dystrophy mutations are restricted to a single codon within this domain.

The retGC1 dimers that form the functional protein interact via a coiled coil formed between the dimerization domains of the two subunits (Ramamurthy et al 2001). Coiled-coils share a characteristic heptad repeat where the residues at position 1 and 4 are hydrophobic (Fig. 4). Hydrophobic interactions between the two α-helices occur every fourth residue of the sequence. This is frequently leucine, thereby giving rise to a hydrophobic core which maintains the structure. The COILS2 program (*http://tofu.tamu.edu/Pise/5.a/coils2.html*) predicts a coiled coil between residues 815–844 involving four heptads, which is broken by Arg838. Replacement of this residue with either Cys, Ser or His increase the probability of the coiled-coil structure continuing for further turns. The effect of this is to increase the stability of the structure such that it resists disruption via the binding of Ca^{2+}-loaded GCAP1.

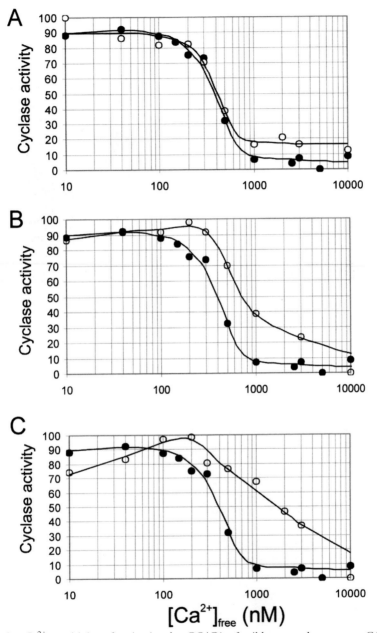

FIG. 3. Ca^{2+} sensitivity of activation by GCAP1 of wild-type and mutant retGC1. (A) Arg838Cys, (B) Arg838His, (C) Arg838Ser. Wild-type retGC1, filled circles; mutant retGC1, open circles. Cyclase activity is expressed as percentage maximal activity. The assay contained 8 μM GCAP1 in all cases.

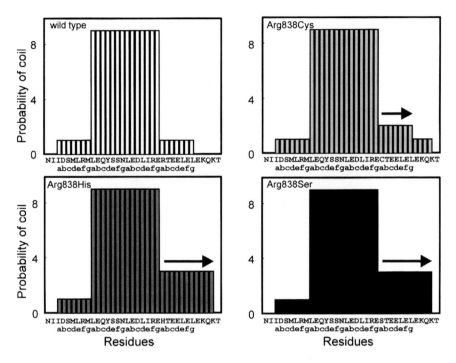

FIG. 4. Probability of forming and extending an α-helical coiled-coil region in dimerization domain of wild-type and mutant retGC1.

Discussion

The major effect of the retGC1 mutations is a reduced sensitivity to suppression of cyclase activity by Ca^{2+}/GCAP1. The altered Ca^{2+} sensitivity is predicted to result in residual (constitutive) activity even at elevated $[Ca^{2+}]_{free}$, with consequent changes in the equilibrium of Ca^{2+} and cGMP concentrations. The three disease-associated mutations are not entirely equivalent, with a shift in the $[Ca^{2+}]_{free}$ for half maximal activation to higher concentrations in the order of Arg838Ser > Arg838His > Arg838Cys. These results parallel the effects of the mutations on the disease phenotype, with patients with the Arg838Ser mutation generally displaying more severe symptoms at an earlier age than those with the Arg838Cys or Arg838His mutations (Downes et al 2001).

Analysis of the sequence of the dimerization domain using the structure prediction program COILS reveals that the residue at position 838 is a key determinant of the extent of the coiled-coil structure responsible for holding together the active retGC1 dimer. Arg838 is predicted to disrupt the structure, limiting it to just four turns of each helix, whilst substitution with other residues

Wild type

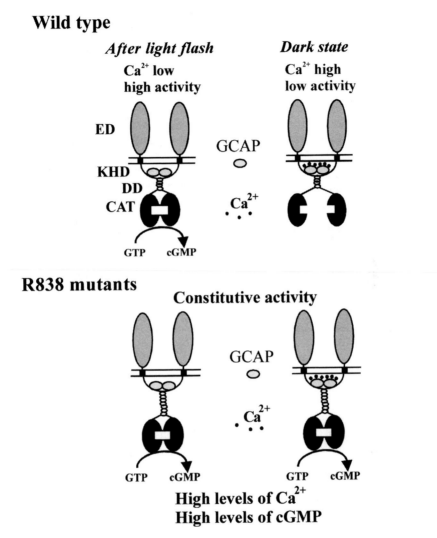

FIG. 5. Model of the regulation of wild-type and mutant retGC1 by GCAP1 under light and dark conditions of illumination. ED, extracellular domain; KHD, kinase homology domain; DD, dimerization domain; CAT, catalytic domain.

results in a higher probability that the structure will continue for further turns. Arg838Ser and Arg838His substitutions might be expected to have the largest impact on structure. The structural consequences of these substitutions has been examined in greater detail by Ramamurthy et al (2001) and a model to account for the effect of Arg838 substitutions is presented in Fig. 5. In the dark when intracellular Ca^{2+} is high, repulsion between Ca^{2+}/GCAP1 monomer units forces the retGC1 dimer apart, inhibiting the cyclase activity. After light stimulation when the $[Ca^{2+}]_{free}$ falls, the GCAP1 dimer facilitates formation of the coiled-coil structure in the dimerization domain of retGC1 and enzyme activation occurs. The effect of substitutions at site 838 is therefore to stabilise the dimeric structure such that it is more resistant to disruption by Ca^{2+}/GCAP1.

The Tyr99Cys and Glu155Gly mutations in GCAP1 also exert their effect via a change in Ca^{2+} sensitivity and the magnitude of the effect is very similar to the more severe retGC1 mutations. The mutations are in EF hands 3 and 4 respectively, the protein domains directly involved in Ca^{2+} binding. In both cases therefore, the number of Ca^{2+} binding sites is reduced from three to two and the reduced level of cyclase inhibition by the mutant GCAP1s at higher $[Ca^{2+}]_{free}$ is therefore not surprising.

Photoreceptors contain two distinct forms of retinal guanylate cyclase, retGC1 and retGC2, and three forms of activating protein, GCAP1, GCAP2 and GCAP3. Various studies have revealed the presence of retGC1 (Liu et al 1994, Hallet et al 1996) and GCAP1 (Cuenca et al 1998) in both cone and rod photoreceptors, so differential expression does not explain the restriction of cell loss to cones in the GCAP1 mutations. LCA1 is caused by recessive mutations in retGC1 (Perrault et al 1996) and the mouse knock-out of the orthologue of human retGC1 results in a distinct retinal pathology (Yang et al 1999), indicating in both cases that retGC2 does not protect photoreceptors from the damaging effect of mutations in retGC1. Likewise, the introduction of a GCAP1 transgene (Howes et al 2002) into a mouse line lacking both GCAP1 and GCAP2 fully restores the normal flash responses whereas Mendez et al (2001) have shown that GCAP2 does not restore the normal physiological response of rod photoreceptors to light flashes. These results indicate that GCAP2 is relatively unimportant in phototransduction. What remains unclear however is why this reduced sensitivity to increasing $[Ca^{2+}]_{free}$ results in a cone dystrophy for the GCAP1 mutations but a cone-rod dystrophy for the retGC1 mutations.

What are the possible consequences of the reduced sensitivity of mutant GCAP1 and retGC1 to $[Ca^{2+}]_{free}$ and how does it relate to the pathology of the cone and cone–rod dystrophies? The operating range of rods is substantially extended by the regulation of the cyclase activity of retGC1 by GCAP1 in decreasing the flash sensitivity in darkness and increasing the incremental flash sensitivity in bright steady light. Reduced Ca^{2+} sensitivity of the mutant systems may serve to reduce this range and thereby account for the photophobia experienced by a number of our cone and cone-rod patients. The net result of the changes in Ca^{2+} sensitivity might be expected to be the maintenance of cGMP levels in the cell above that required to keep the cGMP-gated cation channels open, resulting in persistently high intracellular Ca^{2+} levels. Persistent elevated Ca^{2+} levels in cells tend to disrupt the membrane potential of the mitochondrial outer membrane, leading to release of cytochrome C, with subsequent caspase activation and apoptosis (Green & Reed 1998). Treatment with Ca^{2+} channel blockers like L-*cis*-diltiazem which block both rod and cone cGMP-gated channels may therefore be beneficial. However, it is not possible to predict with any certainty the effect on cellular cGMP and Ca^{2+} concentrations because it is unclear as to what extent the activity of other components of the phototransduction cascade might change in a compensatory manner. The generation of transgenic animals will be especially useful therefore in extending our understanding of this aspect of the aetiology of these disorders and this work is currently in progress.

Acknowledgments

This work was supported by a Programme Grant from the Wellcome Trust.

References

Baylor D 1996 How protons start vision. Proc Natl Acad Sci USA 93:560–565

Cuenca N, Lopez S, Howes K, Kolb H 1998 The localization of guanylyl cyclase-activating proteins in the mammalian retina. Invest Ophthalmol Vis Sci 39:1243–1250

Dizhoor AM, Hurley JB 1999 Regulation of photoreceptor membrane guanylyl cyclases by guanylyl cyclase activator proteins. Methods 19:521–531

Dizhoor AM, Lowe DG, Olshevskaya EV, Laura RP, Hurley JB 1994 The human photoreceptor membrane guanylyl cyclase, RetGC, is present in outer segments and is regulated by calcium and a soluble activator. Neuron 12:1345–1352

Dizhoor AM, Olshevskaya EV, Henzel WJ et al 1995 Cloning, sequencing, and expression of a 24-kDa Ca^{2+}-binding protein activating photoreceptor guanylyl cyclase. J Biol Chem 270:25200–25206

Dizhoor AM, Boikov SG, Olshevskaya EV 1998 Constitutive activation of photoreceptor guanylate cyclase by Y99C mutant of GCAP-1. Possible role in causing human autosomal dominant cone degeneration. J Biol Chem 273:17311–17314

Downes SM, Payne AM, Kelsell RE et al 2001 Autosomal dominant cone-rod dystrophy with mutations in the guanylate cyclase 2D gene encoding retinal guanylate cyclase-1. Archiv Ophthalmol 119:1667–1673

Dryja TP, Rucinski DE, Chen SH, Berson EL 1999 Frequency of mutations in the gene encoding the alpha subunit of rod cGMP-phosphodiesterase in autosomal recessive retinitis pigmentosa. Invest Ophthalmol Vis Sci 40:1859–1865

Gorczyca WA, Gray-Keller MP, Detwiler PB, Palczewski K 1994 Purification and physiological evaluation of a guanylate cyclase activating protein from retinal rods. Proc Natl Acad Sci USA 91:4014–4018

Gorczyca WA, Polans AS, Surgucheva IG, Subbaraya I, Baehr W, Palczewski K 1995 Guanylyl cyclase activating protein. A calcium-sensitive regulator of phototransduction. J Biol Chem 270:22029–22036

Green DR, Reed JC 1998 Mitochondria and apoptosis. Science 281:1309–1312

Haeseleer F, Sokal I, Li N et al 1999 Molecular characterization of a third member of the guanylyl cyclase-activating protein subfamily. J Biol Chem 274:6526–6535

Hallett MA, Delaat JL, Arikawa K, Schlamp CL, Kong F, Williams DS 1996 Distribution of guanylate cyclase within photoreceptor outer segments. J Cell Sci 109:1803–1812

Howes KA, Pennesi ME, Sokal I et al 2002 GCAP1 rescues rod photoreceptor response in GCAP1/GCAP2 knockout mice. EMBO J 21:1545–1554

Imanishi Y, Li N, Sokal I et al 2002 Characterization of retinal guanylate cyclase-activating protein 3 (GCAP3) from zebrafish to man. Eur J Neurosci 15:63–78

Kelsell RE, Gregory-Evans K, Payne AM et al 1998 Mutations in the retinal guanylate cyclase (*RETGC-1*) gene in dominant cone-rod dystrophy. Hum Mol Genet 7:1179–1184

Liu X, Seno K, Nishizawa Y et al 1994 Ultrastructural localization of retinal guanylate cyclase in human and monkey retinas. Exp Eye Res 59:761–768

Lowe DG, Dizhoor AM, Liu K et al 1995 Cloning and expression of a second photoreceptor-specific membrane retina guanylyl cyclase (RetGC), RetGC-2. Proc Natl Acad Sci USA 92:5535–5539

McLaughlin ME, Sandberg MA, Berson EL, Dryja TP 1993 Recessive mutations in the gene encoding the beta-subunit of rod phosphodiesterase in patients with retinitis pigmentosa. Nat Genet 4:130–134

Mendez A, Burns ME, Sokal I et al 2001 Role of guanylate cyclase-activating proteins (GCAPs) in setting the flash sensitivity of rod photoreceptors. Proc Natl Acad Sci USA 98:9948–9953

Newbold RJ, Deery EC, Walker CE et al 2001 The destabilisation of human GCAP1 by a proline to leucine mutation might cause a cone-rod dystrophy. Hum Mol Genet 10:47–54

Otto-Bruc A, Buczylko J, Surgucheva I et al 1997 Functional reconstitution of photoreceptor guanylate cyclase with native and mutant forms of guanylate cyclase-activating protein 1. Biochemistry 36:4295–4302

Payne AM, Downes SM, Bessant DA et al 1998 A mutation in guanylate cyclase activator 1A (GUCA1A) in an autosomal dominant cone dystrophy pedigree mapping to a new locus on chromosome 6p21.1. Hum Mol Genet 7:273–277

Payne AM, Morris AG, Downes SM et al 2001 Clustering of mutations in the retinal guanylate cyclase (*GUCY2D*) gene in patients with dominant cone-rod macular dystrophies. J Med Genet 38:611–614

Palczewski K, Subbaraya I, Gorczyca WA et al 1994 Molecular cloning and characterization of retinal photoreceptor guanylyl cyclase-activating protein. Neuron 13:395–404

Perrault I, Rozet JM, Calvas P et al 1996 Retinal-specific guanylate cyclase gene mutations in Leber's congenital amaurosis. Nat Genet 14:461–464

Pittler SJ, Baehr W 1991 Identification of a nonsense mutation in the rod photoreceptor cGMP phosphodiesterase beta-subunit gene of the rd mouse. Proc Natl Acad Sci USA 88:8322–8326

Polans A, Baehr W, Palczewski K 1996 Turned on by Ca^{2+}! The physiology and pathology of Ca^{2+}-binding proteins in the retina. Trends Neurosci 19:547–554

Pugh EN Jr 1996 Cooperativity in cyclic nucleotide-gated ion channels. J Gen Physiol 107:165–167

Pugh EN Jr, Duda T, Sitaramayya A, Sharma RK 1997 Photoreceptor guanylate cyclases: a review. Biosci Rep 17:429–473

Ramamurthy V, Tucker C, Wilkie SE, Daggett V, Hunt DM, Hurley JB 2001 Interactions within the coiled-coil domain of RetGC-1 guanylyl cyclase are optimized for regulation rather than for high affinity. J Biol Chem 276:26218–26229

Rudnicka-Nawrot M, Surgucheva I, Hulmes JD et al 1998 Changes in biological activity and folding of guanylate cyclase-activating protein 1 as a function of calcium. Biochemistry 37:248–257

Sokal I, Li N, Surgucheva I et al 1998 GCAP1 (Y99C) mutant is constitutively active in autosomal dominant cone dystrophy. Mol Cell 2:129–133

Wilkie SE, Newbold RJ, Deery E et al 2000 Functional characterisation of missense mutations at codon 838 in retinal guanylate cyclase correlates with disease severity in patients with autosomal dominant cone-rod dystrophy. Hum Mol Genet 9:3065–3073

Wilkie SE, Li Y, Deery EC et al 2001 Identification and functional consequences of a new mutation (E155G) in GCAP1 causing autosomal dominant cone dystrophy. Am J Hum Genet 69:471–480

Yang RB, Robinson SW, Xiong WH, Yau KW, Birch DG, Garbers DL 1999 Disruption of a retinal guanylyl cyclase gene leads to cone-specific dystrophy and paradoxical rod behavior. J Neurosci 19:5889–5897

DISCUSSION

Baehr: There are three GCAPs in the human retina. Did you screen all the individuals of families for GCAP2 and GCAP3 mutations?

Hunt: No.

Molday: The knockout mouse for retCG1 shows primarily a cone degeneration phenotype. What is the state of the rod cells in relation to the presence and location of retCG1 and retCG2?

Hunt: The knockout phenotype would indeed seem to fit with the fact that the cones seem to be more sensitive than the rods to loss of retGC1. Bear in mind however that a knockout null mutation of retGC1 is not equivalent to the dominant mutations that we have studied.

Baehr: Yang et al (1999) said that the rod response is abnormal but it is there in the retGC1 knockout. This means that there must be another cyclase in rods in the absence of retGC1. It is a mystery.

Zack: In terms of expression patterns as well as clinical phenotype, are there differences between different cone subtypes?

Hunt: I'm not sure.

Bird: I can't remember, but the photophobia that they have is really quite exceptional. Members in the initial family find even threshold stimuli to be unpleasant. Hence it was very difficult to characterize the disorder. The arginine/cysteine mutation is much milder, and as children the patients carrying this couldn't see in bright light. However, at the age of 40 years they still had good acuity. They had this strange symptom that in bright light they couldn't see, yet they maintained quite good cone function under mesopic conditions for years.

Zack: In terms of gene expression or in the knockout, are there differences between different wavelength cones?

Hunt: As far as I know this hasn't been looked at. I don't think there is any evidence for the differential expression of retGC isoforms in the different cone classes.

Aguirre: One of the questions you raised, which I think is important to resolve, is what is really causing the cells to become sick and die? cGMP is rising, but this might not be what is killing the cells. It may be affecting transduction. The Ca^{2+} seems a more promising possibility, and one that might be more easily addressable for treatment using Ca^{2+} channel blockers. Several years ago we did a study in which we crossed mouse strains with different mutations, including *rd* and *rds*. The *rd* retinas lost photoreceptors very quickly and had a very high level of cGMP. The *rds* mutants lost them slowly and had lower than normal levels of cGMP. The double homozygotes had as high a level as the *rd* mutants, but lost photoreceptors at an intermediate rate. These cells are also abnormal in that they don't have an outer segment. This may be a compartmentalization issue as the double *rd/rds* homozygotes lacked an outer segment.

Farber: How high were the levels of cGMP? Did you measure them?

Aguirre: The levels were measured and in both the *rd* and *rd/rds* double homozygotes the peak levels of cGMP were approximately 10-fold higher than normal controls.

Hunt: This is where the animal models come in. They will be particularly valuable in determining changes in cGMP and Ca^{2+} levels arising as a result of these mutations.

Reference

Yang RB, Robinson SW, Xiong WH, Yau KW, Birch DG, Garbers DL 1999 Disruption of a retinal guanylyl cyclase gene leads to cone-specific dystrophy and paradoxical rod behavior. J Neurosci 19:5889–5897

Isotretinoin treatment inhibits lipofuscin accumulation in a mouse model of recessive Stargardt's macular degeneration[1]

Roxana A. Radu*, Nathan L. Mata*, Steven Nusinowitz*, Xinran Liu†
and Gabriel H. Travis*‡[2]

*Jules Stein Eye Institute, UCLA School of Medicine, Los Angeles, CA 90095, †Center for Basic Neuroscience, UT Southwestern Medical Center, Dallas, TX 75235 and ‡Department of Biological Chemistry, UCLA School of Medicine, Los Angeles, CA 90095, USA

Abstract. Recessive Stargardt's macular degeneration is an inherited blinding disease of children caused by mutations in the *ABCR* gene. The primary pathologic defect in Stargardt's disease is accumulation of toxic lipofuscin pigments such as N-retinylidene-N-retinylethanolamine (A2E) in cells of the retinal pigment epithelium (RPE). This accumulation appears to be responsible for the photoreceptor death and severe visual loss in Stargardt's patients. Here, we tested a novel therapeutic strategy to inhibit lipofuscin accumulation in a mouse model of recessive Stargardt's disease. Isotretinoin (Accutane) has been shown to slow the synthesis of 11-*cis*-retinaldehyde (11*c*RAL) and regeneration of rhodopsin by inhibiting 11-*cis*-retinol dehydrogenase (11*c*RDH) in the visual cycle. Light activation of rhodopsin results in its release of all-*trans*-retinaldehyde (a*t*RAL), which constitutes the first reactant in A2E biosynthesis. Accordingly, we tested the effects of isotretinoin on lipofuscin accumulation in $abcr^{-/-}$ knockout mice. Isotretinoin blocked the formation of A2E biochemically and the accumulation of lipofuscin pigments by electron microscopy. We observed no significant visual loss in treated $abcr^{-/-}$ mice by electroretinography. Isotretinoin also blocked the slower, age-dependent accumulation of lipofuscin in wild-type mice. These results corroborate the proposed mechanism of A2E biogenesis. Further, they suggest that treatment with isotretinoin may inhibit lipofuscin accumulation and thus delay the onset of visual loss in Stargardt's patients. Finally, the results suggest that isotretinoin may be an effective treatment for other forms of retinal or macular degeneration associated with lipofuscin accumulation.

2004 Retinal dystrophies: functional genomics to gene therapy. Wiley, Chichester (Novartis Foundation Symposium 255) p 51–67

---

[1]This paper is reproduced in modified form from Radu RA, Mata NL, Nusinowitz S, Liu X, Sieving PA, Travis GH 2003 Treatment with isotretinoin inhibits lipofuscin accumulation in a mouse model of recessive Stargardt's macular degeneration. Proc Natl Acad Sci USA 100:4742–4747, with permission from the National Academy of Sciences, USA.

[2]This paper was presented at the symposium by Gabriel H. Travis to whom correspondence should be addressed.

Fine central vision in humans is mediated by an area of the retina called the macula, which contains a high density of rod and cone photoreceptors. This region is vulnerable to degeneration in a group of central blinding diseases called the macular degenerations. Age-related macular degeneration, for example, is a common disease of complex aetiology that causes reduced visual acuity and legal blindness in the elderly (Klein et al 1992, Leibowitz et al 1980). Recessive Stargardt's disease is an inherited form of macular degeneration with an onset of central visual loss during childhood (Lee & Heckenlively 1999) and an estimated prevalence of $\sim 1/10\,000$ (Blacharski 1988). Pathologically, Stargardt's is associated with accumulation of fluorescent lipofuscin pigments in cells of the retinal pigment epithelium (RPE) (Birnbach et al 1994, Eagle et al 1980). The RPE is a layer of cells adjacent to the retina that plays a critical role in photoreceptor survival (Steinberg 1985). No effective treatments exist for age-related or Stargardt's macular degeneration.

The gene affected in recessive Stargardt's, *ABCA4* or *ABCR*, encodes a transporter protein in the rims of rod and cone outer-segment discs (Allikmets et al 1997, Azarian & Travis 1997, Illing et al 1997). To study the function of the ABCR protein and the molecular aetiology of Stargardt's, we previously generated mice with a knockout mutation in the *abcr* gene. These animals have a complex

---

FIG. 1.   (*Opposite*) Retinoid pathways in the retina and RPE. (*A*) Visual cycle mediating rhodopsin regeneration. Absorption of a photon (*hv*) by a rhodopsin molecule in a rod outer-segment disc induces photoisomerization of the 11-*cis*-retinaldehyde (11*c*RAL) chromophore, yielding activated metarhodopsin II. After several seconds, metarhodopsin II decays to yield apo-rhodopsin and free all-*trans*-retinaldehyde (a*t*RAL). Elimination of a*t*RAL from the interior of outer-segment discs is facilitated by the ABCR transporter (Weng et al 1999). The a*t*RAL is subsequently reduced to all-*trans*-retinol (a*t*ROL or vitamin A) by all-*trans*-retinol dehydrogenase (a*t*RDH). The a*t*ROL is released from the outer segment and taken up by an RPE cell where it is esterified by lecithin retinol acyl transferase (LRAT) to form an all-*trans*-retinyl ester (a*t*RE). Chemical isomerization is effected by isomerohydrolase (IMH), which uses a*t*RE as a substrate. The resulting 11*c*ROL is oxidized by 11*c*RDH to form 11*c*RAL chromophore. 11*c*RDH is inhibited by isotretinoin with a K_i of $\sim 0.1\,\mu M$ (Gamble et al 2000, Law & Rando 1989). The K_i for inhibition of a*t*RDH by isotretinoin is at least two logs higher (N. L. Mata, R. A. Radu, G. H. Travis, unpublished observations). 11*c*ROL may also serve as a substrate for LRAT to form 11-*cis*-retinyl esters (11*c*RE). The final step is recombination of 11*c*RAL with apo-rhodopsin in the outer segment to form a new molecule of light-sensitive rhodopsin. (*B*) Synthesis of A2E. Following light exposure, newly released a*t*RAL condenses reversibly with phosphatidyl-ethanolamine to form *N*-retinylidene-phosphatidylethanolamine (*N*-ret-PE) (step 1). Rarely, a second molecule of a*t*RAL will condense with *N*-ret-PE to form dihydro-*N*-retinylidene-*N*-retinylphosphatidylethanolamine (A2PE-H$_2$) (step 2). The wavelength of maximal absorption (λ_{max}) for A2PE-H$_2$ is 500 nm. Within the acidic and oxidizing environment of RPE phago-lysosomes, A2PE-H$_2$ is oxidized to *N*-retinylidene-*N*-retinylphosphatidylethanolamine (A2PE) ($\lambda_{max} = 435$ nm) (step 3). Finally, hydrolysis of the phosphate ester yields *N*-retinylidene-*N*-retinylethanolamine (A2E) ($\lambda_{max} = 435$ nm) and phosphatidic acid (step 4) (Mata et al 2000).

ocular phenotype that includes elevated all-*trans*-retinaldehyde (a*t*RAL) following light exposure and accumulation of lipofuscin pigments in cells of the RPE (Mata et al 2001, 2000, Weng et al 1999). These animals also manifest very slow photoreceptor degeneration (Mata et al 2001). On the basis of biochemical analysis of *abcr*$^{-/-}$ mice (Mata et al 2001, 2000, Weng et al 1999) and the results of *in vitro* studies (Ahn et al 2000, Sun et al 1999), ABCR appears to function as an outwardly directed flippase for *N*-retinylidene-phosphatidylethanolamine (*N*-ret-PE), the Schiff-base conjugate of phosphatidylethanolamine and a*t*RAL. Accordingly, ABCR may play a role in the visual cycle for the regeneration of rhodopsin (Fig. 1A) by accelerating removal of a*t*RAL from outer-segment discs (Weng et al 1999).

The major fluorescent component of lipofuscin is the *bis*-retinoid pyridinium salt, *N*-retinylidene-*N*-retinylethanolamine (A2E) (Reinboth et al 1997, Sakai et al 1996). Significant accumulation of A2E is seen in RPE cells from *abcr*$^{-/-}$ mice (Mata et al 2000, Weng et al 1999) and patients with Stargardt's (Birnbach et al

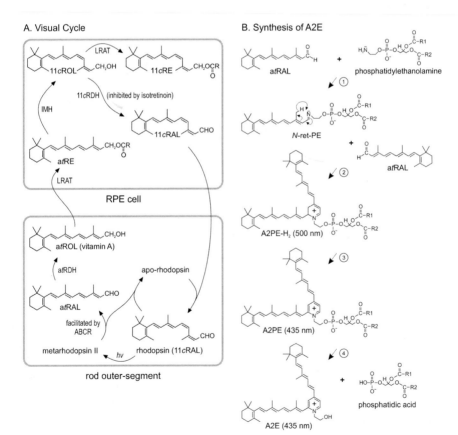

1994, Eagle et al 1980, Mata et al 2000). A2E forms by sequential condensation of a*t*RAL with phosphatidylethanolamine (Fig. 1B) (Mata et al 2000, Parish et al 1998), both of which are elevated in *abcr*$^{-/-}$ retinas (Mata et al 2000, Weng et al 1999). A2E sensitizes RPE cells to light-induced apoptosis (Schutt et al 2000, Sparrow et al 2002, Suter et al 2000) and has an inhibitory effect on phospholipid turnover in RPE phagolysosomes (Finnemann et al 2002). Central blindness in Stargardt's patients results from photoreceptor degeneration caused by A2E-mediated toxicity to the RPE (Sparrow et al 2002, Weng et al 1999). Accordingly, a reasonable strategy to slow the progression of visual loss in Stargardt's patients is to inhibit formation of A2E. Isotretinoin (13-*cis*-retinoic acid or Accutane) is a drug in common use for the treatment of acne (Peck et al 1979). A side-effect of treatment with isotretinoin is reduced night vision (Fraunfelder et al 2001, Weleber et al 1986) due to its inhibitory effect on 11-*cis*-retinol dehydrogenase (11*c*RDH) in RPE cells (Gamble et al 2000, Law & Rando 1989) (Fig. 1A). Treatment of rodents with isotretinoin was shown to delay rhodopsin regeneration and slow recovery of rod sensitivity following light exposure (Sieving et al 2001). Importantly, isotretinoin did not cause photoreceptor degeneration, and actually protected photoreceptors from light-induced damage (Sieving et al 2001). These observations suggest that isotretinoin may safely inhibit formation of A2E in Stargardt's patients by reducing levels of its molecular precursors. In the current study, we investigated this potential treatment strategy in *abcr*$^{-/-}$ mice (Radu et al 2003).

Results

Isotretinoin causes delayed dark adaptation in wild-type and *abcr*$^{-/-}$ mice. All vertebrates experience a period of reduced visual sensitivity following light exposure, called dark adaptation. As described (Sieving et al 2001), wild-type mice show delayed dark adaptation following a single dose of isotretinoin. We tested the effects of isotretinoin in four-month-old wild-type and *abcr*$^{-/-}$ mice by electroretinography (ERG), which records the electrical response of the retina to a light flash. Mice were dark-adapted overnight and administered a single i.p. dose of isotretinoin at 40 mg/kg. One hour after injection, the mice were exposed to light that bleached ∼ 40% of rhodopsin. We analysed mice by ERG immediately before and at different times after the photobleach. In one set of experiments, we used dim probe flashes to elicit the ERGs. Both wild-type and *abcr*$^{-/-}$ mice dark adapted more slowly following isotretinoin administration (Figs 2A,B). Untreated *abcr*$^{-/-}$ mice dark adapted more slowly than wild-type mice, as described (Weng et al 1999). In another set of experiments, we used bright probe flashes to elicit the ERGs. Here, we observed much smaller differences in dark adaptation between treated and untreated mice of both genotypes (Figs 2C,D). Finally, we recorded

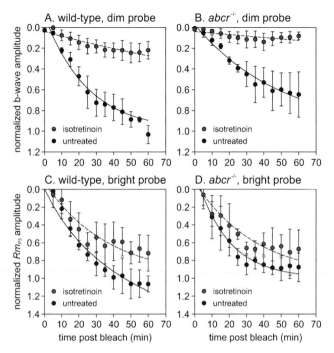

FIG. 2.   Analysis of visual function by electroretinography in wild-type and *abcr*^{−/−} mice after treatment with isotretinoin. (A) *b*-wave amplitudes elicited in wild-type mice with a dim probe flash (−0.91 log scot td-s) at the indicated times following a 40% photobleach divided by the dark-adapted *b*-wave amplitude. The mice received a single injection of isotretinoin (40 mg/kg) one hour before the photobleach (mid-grey circles, upper curve, $n=7$) or received no treatment (black circles, lower curve, $n=4$). (B) Same protocol as in (A) except mice were *abcr*^{−/−} genotype (mid-grey circles, upper curve, $n=4$; black circles, lower curve, $n=5$). (C) Rm_{P3} amplitudes (derived from the leading-edge of the *a*-wave) elicited in wild-type mice with a bright probe flash (3.11 log scot td-s) at the indicated times following a 40% photobleach divided by the dark-adapted Rm_{P3} amplitude. Treatment protocol was similar to (A) (mid-grey circles, $n=7$; black circles, lower curve, $n=7$). (D) Same protocol as in (C) except mice were *abcr*^{−/−} genotype (mid-grey circles, upper curve, $n=9$; black circles, lower curve, $n=9$). Error bars show standard deviations.

ERGs without prior photobleach from four-month-old, dark-adapted wild-type and *abcr*^{−/−} mice that were either untreated, treated with a single i.p. injection of isotretinoin at 40 mg/kg one hour before analysis, or treated for two months with daily i.p. injections of isotretinoin at 20 mg/kg. The dark-adapted ERGs were indistinguishable between treated and untreated mice of both genotypes (data not shown), indicating that isotretinoin did not cause abnormalities in dark-adapted visual function.

Effects of isotretinoin on the levels of visual retinoids. We examined the levels of 11-*cis*-retinaldehyde (11*c*RAL) and 11-*cis*-retinyl esters (11*c*RE) by HPLC in

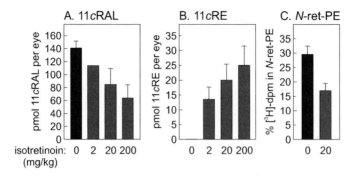

FIG. 3. Effects of isotretinoin on the content of visual retinoids in wild-type and *abcr*$^{-/-}$ eyes. (A) Levels of 11-*cis*-retinaldehyde (11*c*RAL) in light-adapted, wild-type mice that received either no isotretinoin (first bar, $n=3$ mice) or a single injection one hour before tissue collection of isotretinoin at 2.0 ($n=2$), 20 ($n=3$) or 200 ($n=3$) mg/kg. (B) Levels of 11-*cis*-retinyl ester (11*c*RE) in light-adapted wild-type mice that received either no isotretinoin or a single injection of isotretinoin at the indicated doses ($n=3$) as described in (*A*). Note the absence of 11*c*RE in untreated eyes. (C) [^{3}H]-*N*-ret-PE disintegrations per minute (dpm) expressed as a percent of the total [^{3}H]-dpm in phospholipid extracts from eyecups of *abcr*$^{-/-}$ mice that either were untreated (left bar, $n=3$) or treated with isotretinoin at 20 mg/kg per day for 10 days (right bar, $n=4$). Error bars show standard deviations.

eyecups (retina+RPE) from four-month-old, light-adapted wild-type mice one hour following a single i.p. injection of isotretinoin at different doses. As expected, we observed a dose-dependent reduction in 11*c*RAL (Fig. 3A). 11*c*RE was undetectable in eyecups from mice that received no isotretinoin, but was present in treated mice at levels that increased with the dose (Fig. 3B). No significant differences in total retinoid levels were observed between treated and untreated mice. We also looked at formation of *N*-ret-PE in five-month-old *abcr*$^{-/-}$ mice treated with isotretinoin for 10 days by i.p. injection at 20 mg/kg per day. Three days prior to sacrifice, we injected a trace amount (2.0 pmol) of [^{3}H]-a*t*ROL into the vitreous cavity of each eye. After three days under cyclic light, we collected eyecups from the injected eyes and assayed for [^{3}H]-*N*-ret-PE by HPLC with an online radiometric detector. Compared to untreated *abcr*$^{-/-}$ controls, formation of [^{3}H]-*N*-ret-PE was reduced by 40% in the treated eyecups (Fig. 3C).

Sustained treatment with isotretinoin inhibits accumulation of A2PE-H$_2$ and A2E in *abcr*$^{-/-}$ RPE. Reduced formation of *N*-ret-PE in treated retinas suggests that isotretinoin may inhibit accumulation of A2E. To test this possibility, we administered isotretinoin at 40 mg/kg per day via osmotic pump to three-month-old *abcr*$^{-/-}$ mice. After one month, we sacrificed the mice and determined the levels of A2PE-H$_2$ and A2E in RPE by HPLC. In untreated *abcr*$^{-/-}$ mice or mice treated with dimethylsulfoxide (DMSO) vehicle alone, A2PE-H$_2$ increased approximately 2.7-fold in the period between three and four months (Fig. 4A).

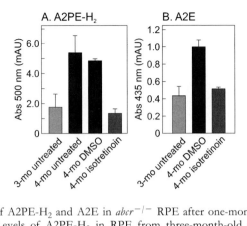

FIG. 4. Levels of A2PE-H$_2$ and A2E in *abcr*$^{-/-}$ RPE after one-month of treatment with isotretinoin. (A) Levels of A2PE-H$_2$ in RPE from three-month-old *abcr*$^{-/-}$ mice before treatment ($n=3$ mice), four-month-old untreated ($n=3$) or DMSO-treated ($n=3$) control mice, and four-month-old mice treated with 40 mg/kg per day isotretinoin ($n=3$). Values expressed as peak-height milliabsorption units (mAU) at wavelength (λ) $= 500$ nm. (B) Levels of A2E in RPE from three-month-old *abcr*$^{-/-}$ mice before treatment ($n=3$), four-month-old DMSO-treated control mice ($n=3$), and four-month-old mice treated with 40 mg/kg per day isotretinoin ($n=3$). Values expressed in mAU at $\lambda = 435$ nm. Error bars show standard deviations.

A2E increased 2.3-fold in DMSO-treated controls during the same period (Fig. 4B). However, the levels of both A2PE-H$_2$ and A2E were virtually unchanged from the levels at three months in mice that received isotretinoin (Figs 4A,B). These data suggest that treatment with isotretinoin suppresses A2PE-H$_2$ and A2E accumulation.

Treatment with isotretinoin inhibits formation of lipofuscin in *abcr*$^{-/-}$ RPE. Next, we looked for an ultrastructural correlate of the reduced A2E. We treated two-month-old *abcr*$^{-/-}$ mice with isotretinoin at 20 mg/kg per day for two months. At the end of the treatment period, we examined the eyes by electron microscopy. Representative electron micrographs of the central retina and RPE from untreated (Fig. 5A) and treated (Fig. 5B) four-month-old *abcr*$^{-/-}$ mice are shown. Oval melanosomes were present at approximately equal density in RPE from both mice. However, the irregularly shaped lipofuscin granules were more abundant in the RPE from untreated mice. These data show that isotretinoin treatment inhibits accumulation of lipofuscin pigments.

Accumulation of A2E in wild-type RPE is also inhibited by isotretinoin. A2E accumulates at slower rates in RPE from wild-type mice and normal humans (Mata et al 2000). To test for a possible effect of isotretinoin on this slow age-dependent accumulation of A2E, we treated two-month-old, wild-type (C57BL/6) mice with daily i.p. injections of isotretinoin at 20 mg/kg for two months. A2E was reduced approximately 40% in these four-month-old mice compared to the levels in

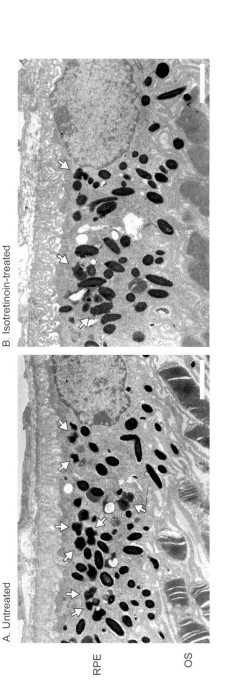

FIG. 5.   Electron microscopic analysis of RPE from four-month-old *abcr*⁻/⁻ mice. (A) RPE and outer segments (OS) from an untreated mouse. (B) RPE and outer segments from a mouse treated with isotretinoin for two months. Grey arrows indicate the irregularly shaped lipofuscin pigment-granules, which are distinct from the larger oval melanosomes. Scale bars represent 2 μm. Note the increased number of lipofuscin granules in RPE from the untreated mouse.

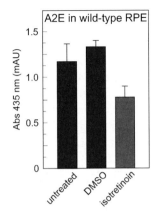

FIG. 6.   Levels of A2E in four-month-old, wild-type mice after two-months of treatment with isotretinoin. A2E levels (mAU at $\lambda=435\,nm$) are shown in RPE from untreated mice ($n=4$ mice), from mice that received daily injections of DMSO vehicle ($n=3$), and from mice that received daily injections of isotretinoin at $20\,mg/kg$ ($n=4$). Error bars show standard deviations.

littermate controls who received either no treatment or injections of DMSO vehicle (Fig. 6). The A2E-absorption data in Fig. 6 represent a sum of peak areas from the different A2E isoforms. Because of the dissimilar light-exposure histories and A2E-quantitation methods, A2E levels in Figs 4 and 6 are not directly comparable. Under controlled conditions, A2E levels are ~10-fold higher in three-month-old *abcr*⁻/⁻ compared to wild-type mice (Mata et al 2000). The data in Fig. 6 show that isotretinoin acts independently of the *abcr* genotype to reduce accumulation of A2E.

Discussion

The primary deficit in *abcr*⁻/⁻ mice and patients with Stargardt's is functional loss of the ABCR transporter. Impaired clearance of N-ret-PE from outer segments that lack ABCR favours the secondary condensation of N-ret-PE with a*t*RAL to form A2PE-H$_2$ (Mata et al 2000) (Fig. 1B). The level of a*t*RAL is approximately 1.5-fold higher in *abcr*⁻/⁻ compared to wild-type outer segments (Weng et al 1999), suggesting that modest reductions in a*t*RAL may be all that is required to inhibit formation of A2PE-H$_2$ and hence A2E. In the current study, we sought to inhibit formation of A2E by reducing production of retinaldehyde in outer segments.

Isotretinoin inhibits 11*c*RDH in RPE cells (Law & Rando 1989, Gamble et al 2000). We analysed the kinetics of dark adaptation in isotretinoin-treated and untreated mice by ERG with bright and dim probe flashes. Dim flashes provide non-saturating stimulation of rods, resulting in an ERG response that is dependent on the level of rhodopsin. Bright flashes saturate the rod response and hence elicit ERGs less sensitive to rhodopsin levels. Both wild-type and $abcr^{-/-}$ mice showed smaller delays in dark adaptation following isotretinoin administration with bright compared to dim probe flashes (Fig. 2). These results suggest that isotretinoin reduced rhodopsin levels in both wild-type and $abcr^{-/-}$ retinas.

We confirmed the inhibitory effect of isotretinoin on 11cRDH (Fig. 1A) by demonstrating dose-dependent reductions in 11cRAL (Fig. 3A). Since 11cROL is esterified by LRAT (Mata & Tsin 1998, Saari & Bredberg 1987), treatment with isotretinoin should result in increased 11cRE, which we observed (Fig. 3B). 11cRE was not detectable in four-month-old untreated mice, as previously described (Driessen et al 2000, Mata et al 2002), presumably due to rapid oxidation of 11cROL by 11cRDH (Fig. 1A). In addition to its effect on 11cRDH, might isotretinoin act by inhibiting transport of 11cRAL between RPE and photoreceptors, since retinoic acid binds interphotoreceptor retinoid-binding protein (Fong et al 1984)? This is unlikely in view of the normal rhodopsin regeneration-kinetics observed in $irbp^{-/-}$ knockout mice (Palczewski et al 1999). N-ret-PE is nearly threefold higher in retinas from $abcr^{-/-}$ compared to wild-type mice due to increased atRAL (Mata et al 2000, Weng et al 1999). Lowering atRAL in $abcr^{-/-}$ photoreceptors should result in reduced N-ret-PE. We injected radiolabelled vitamin A into the vitreous cavities of mice that previously received isotretinoin for one week and measured levels of labelled N-ret-PE. This design permitted us to consider only N-ret-PE formed during the treatment period. We observed an approximate twofold reduction in labelled N-ret-PE (Fig. 3). Since N-ret-PE gives rise to A2E (Fig. 1B), this result suggests that long-term treatment with isotretinoin may suppress A2E accumulation. To test this possibility, we measured levels of $A2PE-H_2$ and A2E in four-month-old $abcr^{-/-}$ RPE after one month of treatment. A2E accumulates rapidly in $abcr^{-/-}$ mice between two and four months of age (Mata et al 2000). In the current study, $A2PE-H_2$ increased almost threefold and A2E approximately doubled in untreated or DMSO-treated $abcr^{-/-}$ control mice (Fig. 4). However, in animals that received isotretinoin, the levels of $A2PE-H_2$ and A2E at four months were virtually unchanged from the starting levels in three-month-old mice (Fig. 4). Thus, treatment with isotretinoin completely suppressed subsequent accumulation of $A2PE-H_2$ and A2E. Since A2E is the major fluorophore of lipofuscin (Reinboth et al 1997, Sakai et al 1996), we also examined the morphological effects of isotretinoin treatment on lipofuscin pigment-formation in $abcr^{-/-}$ eyes. We observed significantly fewer pigment granules in RPE from treated animals compared to untreated controls (Fig. 5). We observed no changes in the ultrastructure of outer segments from treated $abcr^{-/-}$ mice. Finally, we looked for an effect of isotretinoin treatment on the much slower accumulation of A2E in wild-type RPE. After two months of treatment, A2E in RPE from four-month-old, C57BL/6 mice was reduced approximately 40% compared to untreated or DMSO-treated controls (Fig. 6). These data corroborate the proposed pathway for A2E biogenesis (Mata et al 2000, Parish et al 1998) depicted in Fig. 1B. Further, the data establish that treatment with isotretinoin inhibits formation of A2E and lipofuscin in *abcr*$^{-/-}$ and wild-type mice.

The phenotypes in *abcr*$^{-/-}$ mice and humans with Stargardt's are strikingly similar. Both show delayed dark adaptation, elevated phosphatidylethanolamine in retina, elevated A2PE-H$_2$ and A2E in RPE cells, and accumulation of lipofuscin pigments in RPE (Birnbach et al 1994, Eagle et al 1980, Mata et al 2000, Weng et al 1999). Given the physiological, biochemical, morphological, and genetic similarities between *abcr*$^{-/-}$ mice and Stargardt's patients, a treatment that ameliorates the mouse phenotype should have a similar effect in humans. Previously, we showed that A2E synthesis can be virtually blocked by raising *abcr*$^{-/-}$ mice in total darkness (Mata et al 2000). This observation suggests that Stargardt's patients may slow progression of their disease by limiting light exposure. In the current study, we replicated the effects of light deprivation on A2E and lipofuscin accumulation in *abcr*$^{-/-}$ mice by inhibiting rhodopsin regeneration with isotretinoin. This treatment was associated with delayed dark adaptation but normal absolute rod-threshold sensitivity after prolonged dark-adaptation. Delayed dark adaptation has also been reported in humans taking isotretinoin, although the prevalence of this side effect is difficult to quantitate (Fraunfelder et al 2001, Weleber et al 1986). The dose of isotretinoin in the current study (20–40 mg/kg per day) is much higher than the therapeutic dose used in humans to treat acne (0.5–2.0 mg/kg per day). However, the rate of clearance for isotretinoin is 10–20-fold faster in mice than humans (Nau 2001). We observed significant effects on the levels of 11*c*RAL and 11*c*RE following a single injection of isotretinoin at 2.0 mg/kg (Fig. 3A,B), suggesting that this dose inhibits 11*c*RDH. Together, these observations suggest that suppression of A2E and lipofuscin accumulation may be achieved in Stargardt's patients with standard doses of isotretinoin. Finally, isotretinoin or other inhibitors of 11*c*RDH may be effective treatments for retinal or macular degenerations of different aetiologies associated with A2E accumulation.

References

Ahn J, Wong JT, Molday RS 2000 The effect of lipid environment and retinoids on the ATPase activity of ABCR, the photoreceptor ABC transporter responsible for Stargardt macular dystrophy. J Biol Chem 275:20399–20405

Allikmets R, Shroyer NF, Singh N et al 1997 Mutation of the Stargardt disease gene (ABCR) in age-related macular degeneration. Science 277:1805–1807

Azarian SM, Travis GH 1997 The photoreceptor rim protein is an ABC transporter encoded by the gene for recessive Stargardts-disease (ABCR). FEBS Letters 409:247–252

Birnbach CD, Jarvelainen M, Possin DE, Milam AH 1994 Histopathology and immunocytochemistry of the neurosensory retina in fundus flavimaculatus. Ophthalmology 101:1211–1219

Blacharski PA 1988 Fundus flavimaculutus. Raven Press, New York

Driessen C, Winkens HJ, Hoffmann K et al 2000 Disruption of the 11-cis-retinol dehydrogenase gene leads to accumulation of cis-retinols and cis-retinyl esters. Mol Cell Biol 20:4275–4287

Eagle RC Jr, Lucier AC, Bernardino VB Jr, Yanoff M 1980 Retinal pigment epithelial abnormalities in fundus flavimaculatus: a light and electron microscopic study. Ophthalmology 87:1189–1200

Finnemann SC, Leung LW, Rodriguez-Boulan E 2002 The lipofuscin component A2E selectively inhibits phagolysosomal degradation of photoreceptor phospholipid by the retinal pigment epithelium. Proc Natl Acad Sci USA 99:3842–3847

Fong SL, Liou GI, Landers RA, Alvarez RA, Bridges CD 1984 Purification and characterization of a retinol-binding glycoprotein synthesized and secreted by bovine neural retina. J Biol Chem 259:6534–6542

Fraunfelder FT, Fraunfelder FW, Edwards R 2001 Ocular side effects possibly associated with isotretinoin usage. Am J Ophthalmol 132:299–305

Gamble MV, Mata NL, Tsin AT, Mertz JR, Blaner WS 2000 Substrate specificities and 13-cis-retinoic acid inhibition of human, mouse and bovine cis-retinol dehydrogenases. Biochim Biophys Acta 1476:3–8

Illing M, Molday LL, Molday RS 1997 The 220-kDa rim protein of retinal rod outer segments is a member of the ABC transporter superfamily. J Biol Chem 272:10303–10310

Klein R, Klein BE, Linton KLP 1992 Prevalence of age-related maculopathy: The Beaver Dam Eye Study. Ophthalmology 99:933–943

Law WC, Rando RR 1989 The molecular basis of retinoic acid induced night blindness. Biochem Biophys Res Comm 161:825–829

Lee BL, Heckenlively JR 1999 Stargardt's Disease and Fundus flavimaculatus. In: Guyer DR, Yannuzzi LA, Chang S, Shields JA, Green WR (eds) Retina, vitreous, macula. 1st edn, WB Saunders, Darien IL, USA, p 978–988

Leibowitz HM, Krueger DE, Maunder LR et al 1980 The Framingham Eye Study monograph: an ophthalmological and epidemiological study of cataract, glaucoma, diabetic retinopathy, macular degeneration, and visual acuity in a general population of 2631 adults, 1973–1975. Surv Ophthalmol 24:335–610

Mata NL, Tsin AT 1998 Distribution of 11-cis LRAT, 11-cis RD and 11-cis REH in bovine retinal pigment epithelium membranes. Biochim Biophys Acta 1394:16–22

Mata NL, Weng J, Travis GH 2000 biosynthesis of a major lipofuscin fluorophore in mice and humans with ABCR-mediated retinal and macular degeneration. Proc Natl Acad Sci USA 97:7154–7159

Mata NL, Tzekov RT, Liu XR, Weng J, Birch DG, Travis GH 2001 Delayed dark-adaptation and lipofuscin accumulation in abcr[+/−] mice: Implications for involvement of ABCR in age-related macular degeneration. Invest Ophthalmol Vis Sci 42:1685–1690

Mata NL, Radu RA, Clemmons RC, Travis GH 2002 Isomerization and oxidation of vitamin a in cone-dominant retinas. A novel pathway for visual-pigment regeneration in daylight. Neuron 36:69–80

Nau H 2001 Teratogenicity of isotretinoin revisited: species variation and the role of all-trans-retinoic acid. J Am Acad Dermatol 45:S183–S187

Palczewski K, Van Hooser JP, Garwin GG, Chen J, Liou GI, Saari JC 1999 Kinetics of visual pigment regeneration in excised mouse eyes and in mice with a targeted disruption of the gene encoding interphotoreceptor retinoid-binding protein or arrestin. Biochemistry 38:12012–12019

Parish CA, Hashimoto M, Nakanishi K, Dillon J, Sparrow J 1998 Isolation and one-step preparation of A2E and iso-A2E, fluorophores from human retinal pigment epithelium. Proc Natl Acad Sci USA 95:14609–14613

Peck GL, Olsen TG, Yoder FW et al 1979 Prolonged remissions of cystic and conglobate acne with 13-cis-retinoic acid. N Engl J Med 300:329–333

Radu RA, Mata NL, Nusinowitz S, Liu X, Sieving PA, Travis GH 2003 Treatment with isotretinoin inhibits lipofuscin accumulation in a mouse model of recessive Stargardt's macular degeneration. Proc Natl Acad Sci USA 100:4742–4747

Reinboth JJ, Gautschi K, Munz K, Eldred GE, Reme CE 1997 Lipofuscin in the retina: quantitative assay for an unprecedented autofluorescent compound (pyridinium bis-retinoid, A2-E) of ocular age pigment. Exp Eye Res 65:639–643

Saari JC, Bredberg DL 1987 Photochemistry and stereoselectivity of cellular retinaldehyde-binding protein from bovine retina. J Biol Chem 262:7618–7622

Sakai N, Decatur J, Nakanishi K, Eldred GE 1996 Ocular age pigment 'A2-E': an unprecedented pyridinium bisretinoid. J Am Chem Soc 118:1559–1560

Schutt F, Davies S, Kopitz J, Holz FG, Boulton ME 2000 Photodamage to human RPE cells by A2-E, a retinoid component of lipofuscin. Invest Ophthalmol Vis Sci 41:2303–2308

Sieving PA, Chaudhry P, Kondo M et al 2001 Inhibition of the visual cycle in vivo by 13-cis retinoic acid protects from light damage and provides a mechanism for night blindness in isotretinoin therapy. Proc Natl Acad Sci USA 98:1835–1840

Sparrow JR, Zhou J, Ben-Shabat S, Vollmer H, Itagaki Y, Nakanishi K 2002 Involvement of oxidative mechanisms in blue-light-induced damage to A2E-laden RPE. Invest Ophthalmol Vis Sci 43:1222–1227

Steinberg RH 1985 Interactions between the retinal pigment epithelium and the neural retina. Doc Ophthalmol 60:327–346

Sun H, Molday RS, Nathans J 1999 Retinal stimulates ATP hydrolysis by purified and reconstituted ABCR, the photoreceptor-specific ATP-binding cassette transporter responsible for Stargardt disease. J Biol Chem 274:8269–8281

Suter M, Reme C, Grimm C et al 2000 Age-related macular degeneration — the lipofuscin component N-retinyl-N-retinylidene ethanolamine detaches proapoptotic proteins from mitochondria and induces apoptosis in mammalian retinal pigment epithelial cells. J Biol Chem 275:39625–39630

Weleber RG, Denman ST, Hanifin JM, Cunningham WJ 1986 Abnormal retinal function associated with isotretinoin therapy for acne. Arch Ophthalmol 104:831–837

Weng J, Mata NL, Azarian SM, Tzekov RT, Birch DG, Travis GH 1999 Insights into the function of Rim protein in photoreceptors and etiology of Stargardt's disease from the phenotype in abcr knockout mice. Cell 98:13–23

DISCUSSION

Bok: Is the dose of isotretinoin you use comparable to those given to patients with cystic acne?

Travis: No, we are using much higher concentrations than those used to treat acne. That said, I am not sure that this is too big a problem. Mice metabolize drugs much more rapidly. If you wanted to put a mouse to sleep with phenobarbital you'd need a dose that would kill humans on a mg/kg body weight basis. The highest dose of isotretinoin for humans is 2 mg/kg, and we are using 20–40 mg/kg in the mice. What dose do we need in humans? If you are getting night blindness with treatment — which is a common side-effect of Accutane treatment — then you are probably influencing the visual cycle, so we probably need doses that are going to cause night blindness. This effect is seen at 1–2 mg/kg in humans.

Ali: Do you see some damage to photoreceptor cells? You didn't mention retinal degeneration *per se*, and I assume that this is because there isn't enough time for this to develop.

Travis: That is an interesting question, and one which we are struggling with. We see very slow photoreceptor degeneration in *abcr* knockout mice. It is one of the most leisurely degenerations that I know of. We have not yet figured out a way to accelerate the rate of degeneration. This has general importance in terms of mouse models. We have always assumed that the rate of degeneration in a mouse (or any animal) model has an implicit denominator, which is the lifespan of the animal. If Rp shows up in the second decade of life in humans, then we'd expect to see it about six weeks in mice. This is pretty consistent, but there is no real reason why it should be true. This may hold in disease processes that are cell autonomous, where you have a P23H mutation in your rhodopsin, for example, and this is killing the photoreceptor that contains rhodopsin. But with *abcr* mutations it is a much more indirect effect. I think it probably has to do with how many photoisomerization cycles have been run. It may be that we don't get significant photoreceptor degeneration in the mouse until it is much older.

Ali: Is there a differential effect on cone and rod degeneration?

Travis: No, we haven't looked at this carefully.

Bok: There is another thing I would urge you to look at. The RPE of the mouse stores much more retinoid than the RPE of the rat. This could very well have an effect.

Travis: Yes, these smaller effects could be significant. We are now trying to run out the clock without killing photoreceptors by putting the *abcr* mutation onto a transducin-null background. This will protect them from light damage through the transduction cascade pathway, so we should be able to expose *abcr* knockout mice to fairly high levels of cyclic light, and hopefully accumulate enough A2E to get photoreceptor degeneration in a few months.

Aguirre: Are the mice more sensitive to light damage as they accumulate more A2E in the RPE?

Travis: They are not dramatically more sensitive.

Hauswirth: Do the old transgenic mice lose RPE cells?

Travis: No. We see a lot of junk, thickening of the membrane and widening of the cells. The melanosomes that are normally distributed more apically become more dispersed into the cell body. However, if you count nuclei in the RPE, the $abcr^{-/-}$ mice are similar to the wild-type mice.

Hauswirth: It would be possible to inject VEGF into an old knockout mouse and induce choroidal vascularization as a measure to see whether Bruch's membrane is breaking down.

Bird: In Stargardt disease in human the full-field ERG remains normal throughout life in 80% of patients. The fact that there isn't widespread photoreceptor loss in the mouse model shouldn't really bother you. In 80% the loss of photoreceptor cells is limited to the macula.

Travis: The macula is clearly a more vulnerable region than the peripheral retina, and we have one big peripheral retina in the mouse.

Swaroop: Have you looked at the secondary effects of A2E accumulation on RPE function? For example, have you looked at the expression of RPE65, bestrophin or other RPE genes late in life of these mice? Could it be that accumulation of A2E is affecting the overall function of RPE?

Travis: We haven't done that, although it would be easy enough to do.

Cremers: I wanted to point out that in humans, RP is the knockout equivalent of your mice, not Stargardt disease. In patients with RP and ABCR null mutations there is a massive pan-retinal degeneration.

Dryja: If you were to start a clinical trial on the basis of this work, would you give the children with ABCR mutations Accutane, or sunglasses, or both?

Travis: We know that if we raise them under total darkness they don't accumulate A2E, so we should certainly limit light. In terms of giving Accutane, it is a reasonable idea. The problem is that Accutane is not a very good drug. We want to inhibit 11-*cis*-retinal dehydrogenase, and there will undoubtedly be better ways of doing it than Accutane. The problem is that the manufacturer of Accutane has relatively little interest in this because this year they lose their patent on the drug.

Hauswirth: On the other side of the coin there are people who have been on Accutane for as long as 10 years or more, so there are a lot of clinical data there.

Bird: When we heard about this we immediately consulted dermatologists. It seems that the human dose in standard use would be difficult to defend for the treatment of Stargardt disease. If the effect can be achieved by levels one-tenth of the human dose then it might be worth looking at. The dermatologists also made the point that there are many other drugs which might well have a similar effect and which would be safer. It would be good to go through these alternative drugs to see if one is safer. I think it would be easy to verify pharmacological effect in humans: you just have record dark adaptation or scotopic sensitivity.

Bok: It would be relatively easy for a drug company to test alternatives because the gene has been cloned, and it could be put into 293 cells.

Zack: You mentioned the two human retinas you have looked at from patients with Stargardt disease. Have you used the same retinoid conjugation assay to look at retinas from age-related macular degeneration (AMD) patients, and particularly ones with putative ABCR mutations?

Travis: Greg Hageman sent me some AMD retina tissue samples. We saw a lot of variability. In some cases we saw A2E levels above controls, and in others they were below age-matched controls. I guess if the RPE is gone, then the place where A2E is found may also be gone. I don't know how to interpret this.

Zack: In terms of AMD, how appropriate do you think the knockout mice are as a model?

Travis: I don't want to get into the whole issue of *ABCR* and AMD. What we observed is a clear phenotype in the *abcr*$^{+/-}$ heterozygotes. If you were to assign a number to the phenotype, it is something like 40% in terms of A2E and the delay in dark adaptation. If anything the lipofuscin seems to be more than 50%. They are not normal. On the basis of this I think there would be something abnormal about a human *ABCR* heterozygous-null. Beyond this I can't say anything.

Bird: In some forms of AMD there are focal increases in the autofluorescence in the pigment epithelium in the early stages of disease, but in many there aren't. Since AMD includes several disorders, you would have to go through a large number of AMD patients, and you would see a big spread. One of the attractions of using drug treatment for this is that I think clinically it might be able to identify an effect. If the focal increases in autofluorescence resolved without there being atrophy, you would say this is unique and is not what happens in disease. If you extended the treatment to AMD you could do exactly the same thing: ask whether focal increases in autofluorescence resolve without atrophy.

Travis: I don't think there is going to be a resolution; I think what may happen is simply stopping progression.

Bird: As soon as reduction in photoreceptor cells occurs autofluorescence disappears in human. It is not known whether it is cleared from the cell by being expelled, or whether it is degraded over a long period. The pigment epithelium appears to be able to rid itself of autofluorescent material.

McInnes: Do those cells die?

Bird: I can't say how it happens. All I am saying is that the pigment epithelium can rid itself of this material. Greg Hageman has shown that the material can be externalized from the cell, either inwards or outwards (personal communication).

Travis: We need to get AMD patients who are actively showing autofluorescence. If we could get postmortem material and assay for A2E, we might be able to identify a subset of AMD patients who have A2E. These may then be good candidates for this treatment.

Bird: They can certainly be identified clinically.

McInnes: This is slightly off-topic, but in western developed countries we all eat a huge excess of vitamin A. Is there any possibility that AMD is a vitamin A ingestion disease in people with the wrong alleles? Or are the rate-limiting steps not there?

Travis: I am just thinking of Stargardt's, which is cleaner. If you were to give a lot of vitamin A to an *abcr*$^{-/-}$ mouse, would that make them worse? They would tend to accumulate esters in the RPE. It may be that the combination of these esters and then extra vitamin A would have deleterious consequences.

Bok: If I were a recessive Stargardt syndrome patient I wouldn't be popping vitamin A pills. In this case, less vitamin A would be better than more vitamin A because there is a disposal problem for one of the vitamin A by-products.

Travis: The question is, would it help patients to put them into a mild vitamin A deficiency state?

Dryja: Isn't β carotene one of the agents that they gave in the AMD study in order to reduce the incidence of AMD?

Bok: Yes. They didn't have access to lutein in sufficient quantities at the time. One of the investigators involved in that study told me that they probably would never have used β carotene if lutein had been available.

Bird: Just as an aside, two or three of our Stargardt's patients have looked at the possibility of going on low vitamin A diets. When they went round the shops trying to find things that don't contain vitamin A it was a difficult search. It is added to so many different foodstuffs.

Thompson: Since the RPE is phagocytosing the outer segments containing this material, in terms of the idea of protecting the RPE against toxicity, one of the strategies might also be to slow down phagocytosis. There now exists the *Mertk* knockout mouse, that is the equivalent to the RCS rat, and defective in RPE phagocytosis. Have you considered putting your *abcr* knockout mouse on a *Mertk*$^{+/-}$ background to see whether slowing down phagocytosis helps reduce the level of A2E?

Travis: No.

Bok: I think that would probably make it worse, because it would increase the sojourn of a disk in the outer segment. That is not what you want. You don't want them hanging around exposed to light and producing the precursor to A2E.

Travis: Is there a partial phenotype of the heterozygous-null *Mertk* mouse?

Thompson: No, not that I know of.

Sahel: In the treated eye, it looked like there might be more lipid droplets in the RPE as opposed to the untreated eye. This is sometimes seen in the RPE65 mutants.

Travis: In the RPE65 mutant there is huge accumulation of retinal esters, and here we don't see that. There is some increase of retinal esters in the treated eyes, but nothing like in the RPE65.

Sahel: I was just concerned that blocking the metabolism would increase the load in the RPE cells.

Travis: In RPE65 there is a real block: there is minimal *cis*-retinoid formed. Here, it is just slowed down, and there is perhaps a twofold increase in esters.

Zack: Have you tried any *in vivo* measurement of autofluorescent material?

Travis: No, I don't really know how to do that.

The expanding roles of *ABCA4* and *CRB1* in inherited blindness

F. P. M. Cremers, A. Maugeri, A. I. den Hollander and C. B. Hoyng*

*Departments of Human Genetics and *Ophthalmology, University Medical Centre Nijmegen, PO Box 9101, 6500 HB Nijmegen, The Netherlands*

Abstract. Mutations in the *ABCA4* gene cause Stargardt disease (STGD), most cases with autosomal recessive (ar) cone-rod dystrophy (CRD), and some cases with atypical ar retinitis pigmentosa (arRP). We found compound heterozygous *ABCA4* mutations in two unrelated patients with STGD and homozygous splice site mutations in their 2nd and 4th degree cousins with RP. Some *ABCA4* mutations display strong founder effects. In Dutch and German STGD patients, the 768G > T mutation is present in 8% and 0.6% of *ABCA4* alleles respectively. Vice versa, the complex L541P;A1038V allele is found in 7% of *ABCA4* alleles in German STGD patients but absent in Dutch patients. As ∼70% of *ABCA4* mutations are known, a microarray-based analysis of known *ABCA4* gene variants allows routine DNA diagnostics in Caucasian patients. Mutations in the *CRB1* gene underlie RP12, some cases with classic arRP, 55% of cases with RP and Coats-like exudative vasculopathy, and 13% of patients with Leber congenital amaurosis (LCA), rendering *CRB1* a significant cause of autosomal recessive retinal dystrophy. Different combinations of mutations in *ABCA4* or *CRB1* can be correlated with disease severity, suggesting that small increments of protein activities in patients might have significant therapeutic effects. Mouse and *Drosophila* studies strongly suggest that both patient groups might benefit from reduced light exposure and therefore should be detected as early as possible using molecular techniques.

2004 Retinal dystrophies: functional genomics to gene therapy. Wiley, Chichester (Novartis Foundation Symposium 255) p 68–84

The retina-specific ATP binding cassette transporter gene (*ABCA4*) encodes the ABCR (RmP) protein located in the membrane of the rod and cone outer segment disks (Allikmets et al 1997, Molday et al 2000). ABCR is thought to function as an outwardly directed flippase for N-retinylidene-phosphatidylethanolamine transporting all-*trans* retinal from the outer segments disk lumen to the photoreceptor cell cytoplasm (Sun et al 1999, Weng et al 1999). Mutations in the *ABCA4* gene are responsible for Stargardt disease (STGD) (Allikmets et al 1997). *ABCA4* null mutations are associated with atypical arRP (Cremers et al 1998, Martinez-Mir et al 1998) and combinations of null and moderately severe mutations are the major cause of arCRD (Maugeri et al 2000, Papaioannou et al

2000). Furthermore, it has been demonstrated that at least two mutations in the *ABCA4* gene are associated with age-related macular degeneration (AMD) at a statistically significant level (Allikmets 2000, and refs therein).

Previously, we described mutations in the *CRB1* gene in a severe, autosomal recessive form of RP (RP12; den Hollander et al 1999). CRB1 is homologous to *Drosophila* Crumbs and contains 19 EGF-like domains, 3 laminin A G-like domains, a transmembrane domain and a 37-amino acid cytoplasmic tail (Fig. 3). In *Drosophila*, the cytoplasmic part of Crumbs is required to maintain the integrity of the zonula adherens in photoreceptor cells, and is a central regulator in the biogenesis of the rhabdomere stalk, a structure that is similar to the inner segments of vertebrate photoreceptors (Izaddoost et al 2002). The extracellular portion of Crumbs is essential to suppress light-induced programmed cell death (Johnson et al 2002). The role of CRB1 in vertebrate photoreceptors may be similar to that of Crumbs, since Crb1 is localized in the inner segments of mouse rods and cones, in the vicinity of the zonula adherens. Interestingly, mouse Crb1 is also associated with the plasma membrane of the cone outer segments (Pellika et al 2002). RP12 is a specific form of RP, in most cases characterized by a preserved para-arteriolar retinal pigment epithelium (PPRPE). Other features of this type of RP are high hyperopia, nystagmus, optic nerve head drusen, vascular sheathing and maculopathy (Heckenlively 1982). Due to early macular involvement, patients suffer from severe visual impairment before the age of 20 years. Mutations in the *CRB1* gene have now been identified in 13 cases with RP12 (den Hollander et al 1999, 2001).

Here, we study the occurrence of *ABCA4* and *CRB1* mutations in patients with several clinically diverse retinal dystrophies, and discuss the prospects of routine *ABCA4* mutation analysis using microarray analysis.

Results and discussion

ABCA4 mutations in families with various types of retinal dystrophy

In the genotype–phenotype model proposed for *ABCA4* (Cremers et al 1998, Maugeri et al 1999, Shroyer et al 1999) (Fig. 1A) there is an inverse relationship between the residual ABCR activity and the severity of the retinal dystrophy. Since different combinations of *ABCA4* mutations lead to different phenotypes, this model implicitly predicts the occurrence of families harbouring different types of retinal disorders caused by different combinations of mutations in the *ABCA4* gene. This was previously shown in a family with arRP and arCRD (Cremers et al 1998) and three families with arRP and STGD (Rozet et al 1999, Rudolph et al 2002, Shroyer et al 2001). In another Dutch family, a female with STGD carries the mild 2588G > C and the severe 768G > T mutation. A second cousin suffers

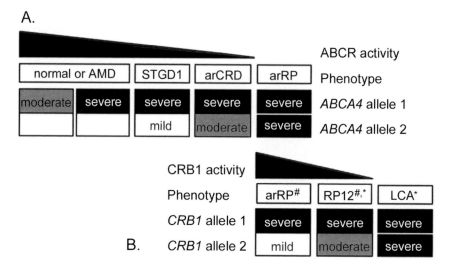

FIG. 1. Genotype–phenotype correlation models for ABCR (A) and CRB1 (B). The ABCR model is corroborated by mutation (Maugeri et al 1999 and references therein) and functional analysis (Sun et al 2000). Based on our findings in patients with classic arRP with Coats-like exudative vasculopathy, RP12 (with or without the Coats' complication), and LCA, we hypothesize that combinations of *CRB1* mutations yielding a high residual activity of CRB1, are associated with classic arRP. Some patients show [#]Coats-like exudative vasculopathy or *preserved para-arteriolar retinal pigment epithelium.

from RP and is homozygous for the 768G > T mutation. His father, a carrier of the 768G > T mutation, shows AMD. In a second family two patients with RP carry a severe splice site mutation (IVS33+1G > A) homozygously. Two grandchildren from one of the patients show STGD and are compound heterozygous for this splice site mutation and the 2588G > C mutation.

Strong founder effects for the ABCA4 mutations
2588G > C, 768G > T and L541P;A1038V

The majority of pathologic variants found in the *ABCA4* gene are very rare. Some alleles are found recurrently in Europe and the USA; some only in specific populations. The 2588G > C variant is by far the most frequent allele in the Netherlands, and was found heterozygously in 35% of patients with STGD (Maugeri et al 1999, F. P. M. Cremers, A. Maugeri, unpublished data) and in 2.9% of healthy individuals from the Netherlands. This variant was not detected in healthy individuals in Portugal, was rare in Spain and France, and increasingly

TABLE 1   Percentage of *ABCA4* alleles in STGD patients from different countries

ABCA4 mutation	*Netherlands*[a,b]	*Germany*[c,d]	*United States*[e,f]
768G > T	7.9% (10/126)	0.6% (2/320)	1.0% (3/290)
L541P;A1038V	0% (0/126)	6.9% (22/320)	3.1% (9/290)
2588G > C	17.5% (22/126)	6.6% (21/320)	3.7% (11/300)
IVS38-10T > C	4.8% (6/126)	3.8% (12/320)	7.9% (23/290)
G1961E	3.2% (4/126)	10.9% (35/320)	5.3% (16/300)

[a]Maugeri et al 1999; [b]F. P. M. Cremers, A. Maugeri, unpublished data; [c]Rivera et al 2000; [d]B. Weber, personal communication ; [e]Lewis et al 1999; [f]R. Allikmets, personal communication.

frequent towards the North West of Europe. Sweden shows a strikingly high carrier frequency of 1 in 18 for this mutation alone (Maugeri et al 2002).

Even more striking allele frequency differences in patients with STGD are observed for two other variants. The 768G > T variant is present in 7.9% and 0.6% of *ABCA4* alleles in Dutch and German STGD patients, respectively, a 13-fold difference (Table 1). On the contrary, the complex L542P;A1038V mutation is absent in Dutch STGD patients, but present in 6.9% of *ABCA4* alleles in German STGD patients (Rivera et al 2000, and B. Weber, personal communication). This complex allele was by far the most frequent *ABCA4* allele among German CRD patients (7 of 19 *ABCA4* mutations in 10 patients) (Maugeri et al 2000, F. P. M. Cremers, A. Maugeri, unpublished results). These data suggest that the 768G > T and L541P;A1038V mutations have arisen relatively recently and have not yet had the time to spread into neighbouring countries.

Cost-effective DNA diagnostics for ABCA4: microarray based analysis

Based on the estimated incidence of mild *ABCA4* alleles in the Western European population and the frequency of families with pseudo-dominantly (8/178) inherited *ABCA4* mutations, an overall heterozygote frequency for *ABCA4* mutations of 4.5–5.8% was estimated (Maugeri et al 1999, Yatsenko et al 2001). In view of the high aggregate frequency of *ABCA4*-associated retinal dystrophy ($\sim 1/8,000$) it is important to design an efficient and cost-effective mutation screening.

Many *ABCA4* missense mutations are so rare that pathogenicity can only be proven using functional assays (Sun et al 2000). Moreover, some polymorphic variants (e.g. IVS38-10T > C) are in linkage disequilibrium with as yet unknown pathologic mutations, complicating DNA diagnostics. In Dutch and German STGD patients, the analysis of four of the five alleles depicted in Table 1 detects

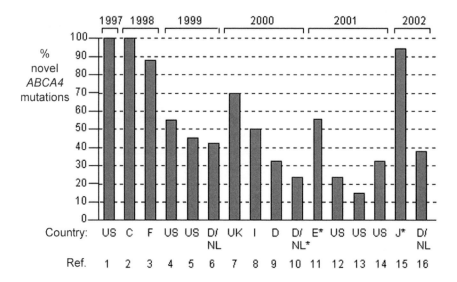

FIG. 2.   Chronological representation of percentages of novel *ABCA4* mutations identified in STGD or CRD patients. Asterisk indicates studies in which less than 25 patients were investigated. The 8 variants in 4 complex alleles encountered in the latter study were calculated as individual mutations. References: (1) (Allikmets et al 1997), (2) (Nasonkin et al 1998), (3) (Rozet et al 1998), (4) (Fishman et al 1999), (5) (Lewis et al 1999), (6) (Maugeri et al 1999), (7) (Papaioannou et al 2000), (8) (Simonelli et al 2000), (9) (Rivera et al 2000), (10) (Maugeri et al 2000), (11) (Paloma et al 2001), (12) (Webster et al 2001), (13) (Yatsenko et al 2001), (14) (Briggs et al 2001), (15) (Fukui et al 2002), (16) A. Maugeri, F. P. M. Cremers, unpublished data. In total, 1212 STGD and 30 CRD patients were analysed.

∼30% of all expected mutations (Table 1). Other mutations are each found in less than 2% of *ABCA4* alleles. Sequence analysis of the *ABCA4* gene is the most sensitive mutation detection technique, but is too expensive (∼2000 USD/DNA sample) for routine DNA diagnostics. It is clear that another technique is needed to implement routine DNA diagnosis for this formidable gene. Microarray based minisequencing can detect the known 350 gene variants in a much faster and cheaper way (Allikmets et al 2001). The efficiency of this method critically depends on the fraction of known mutations. In Fig. 2, the percentage of novel mutations that was discovered in STGD or CRD patients over the last 5 years is depicted chronologically. For each study, the percentage of new mutations was calculated by comparison with the known mutations previously published. Due to the screening of a large number of STGD patients (743) ∼80% of all variants are currently known in the USA (Webster et al 2001, Yatsenko et al 2001, Briggs et al 2001). In Europe a smaller number of STGD patients (361) has been analysed,

TABLE 2   The expected efficiency of CHIP technology

Country	Known mutations	Expected *ABCA4 mutations using CHIP analysis*		
		2 alleles	*1 allele*	*0 alleles*
USA	72%[a]	52%	40%	8%
NL/D	60%	36%	48%	16%
UK/E	35%	10%	43%	47%

[a]Calculated as follows: assuming that 90% of the mutations can be found with currently employed mutation analyses techniques. In the US, 80% of these alleles (see Fig. 2) have now been identified, resulting in an overall known mutation frequency of 72%.

providing an explanation for the lower percentage (65%) of known mutations in the more recent European studies (Rivera et al 2000, F. P. M. Cremers, A. Maugeri, unpublished results). In as yet not well studied or ethnically diverse populations, as e.g. Spain and the UK, this percentage is lower. In a Japanese study of 10 STGD patients, 13 out of 14 mutations were novel, illustrating the variation in occurrence of *ABCA4* mutations in different human populations (Fukui et al 2002). Assuming that the microarray analysis reveals all alleles reliably, we can calculate the expected efficiency of CHIP technology (Table 2). In the USA, both mutations will be found in approximately half of the STGD patients; 1 allele in 40% and no allele in 8% of patients. In the North West of Europe approximately one-third of patients will show both mutations, half of the patients one allele, and 16% no mutation. In less well-studied populations, like the UK and Spain, both mutations will be found in a relatively small fraction of patients. In non-Caucasian populations, a comprehensive mutation analysis is still needed to reach the same effectiveness for microarray based mutation analysis. Nevertheless, for the US and Western European populations, microarray analysis in the future will be the method of choice for *ABCA4* analysis.

The CRB1 gene is involved in 13% of cases with LCA

Due to the early onset of symptoms in patients with RP and PPRPE, we considered *CRB1* to be a good candidate gene for LCA. We identified mutations in seven of 52 (13%) unrelated LCA patients analysed. In six patients we identified both *CRB1* alleles, and in one patient we identified one allele (Fig. 3). Seven out of 13 alleles identified in LCA patients are nonsense, frameshift or splice site mutations, which is a higher fraction than that found for *CRB1* alleles of RP12 (4/26; den Hollander et al 1999, den Hollander et al 2001). In three patients we identified null mutations on both *CRB1* alleles, which suggests that LCA is the most severe phenotype that can be associated with mutations in *CRB1*.

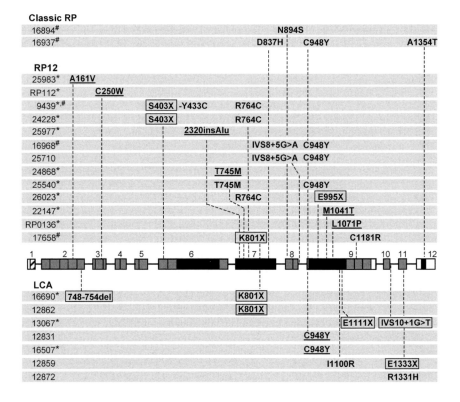

FIG. 3.   Mutation spectrum of *CRB1* gene. Exons are drawn to scale; introns are not. Homozygous mutations are underlined; null alleles are boxed. Exon 1 encodes the signal peptide, the 3 large black boxes represent the laminin A G-like domains, the grey boxes the EGF-like domains. Exon 12 encodes the transmembrane region (black box) and the 37 amino acid cytoplasmic domain. S403X and Y433C as well as D837H and A1354T represent complex *CRB1* alleles. In all patients except 12872 and 16894, both mutant alleles were identified. Clinical features: *Para-arteriolar preservation of the RPE; #Coats-like exudative vasculopathy.

Mutations in *GUCY2D*, *RPE65*, *AIPL1*, *RPGRIP1* and *CRX* account for 6–20%, 3–16%, 7%, 6%, 3% and 1% of LCA cases, respectively (Cremers et al 2002 and references therein). We and others identified *CRB1* mutations in 13% and 9% of patients with LCA, respectively (den Hollander et al 2001, Lotery et al 2001), suggesting that *CRB1* mutations are a significant cause of LCA.

The CRB1 gene is involved in RP with or without Coats-like exudative vasculopathy

Coats-like exudative vasculopathy is a relatively rare complication of RP, which can develop in later stages of the disease and is characterized by vascular abnormalities, yellow extravascular lipid depositions and retinal detachment. Coats-like exudative vasculopathy is observed in 1–3% of RP patients (Khan et al 1988 and references therein), and RP12 patients are at an increased risk to develop Coats-like changes (van den Born et al 1994). To determine whether *CRB1* mutations could be a common cause of RP with Coats-like exudative vasculopathy, we ascertained nine RP patients with this complication. We identified *CRB1* mutations in five patients; in two patients (9439, 16937) we identified three sequence changes, in two patients (16968, 17658) we found compound heterozygous mutations and in one patient (16894) we identified one allele (Fig. 3) (den Hollander et al 2001, and A. I. den Hollander, F. P. M. Cremers, unpublished data). Three of these patients had a severe form of RP (9439, 16968, 17658) with characteristics of RP12, while two patients (16894, 16937) had a later onset classic form of RP.

In the families of five probands with RP and Coats-like exudative vasculopathy, 10 patients were affected with RP, and two of them had not developed the Coats-like complication (den Hollander et al 2001). Among the RP12 patients, we observed combinations of the same mutations in two patient pairs. Patients 9439 and 24228 both carry S403X and R764C mutations and show the PPRPE picture, while only 9439 suffers from the Coats-like complication. Patients 16968 and 25710 are both compound heterozygous for the IVS8+5G > A and C948Y mutations and both do not show the PPRPE picture, while only patient 16968 shows the Coats-like complication. These findings, together with the observation that all patients but one had developed the complication unilaterally, strengthen the idea that *CRB1* mutations are an important risk factor for the development of this severe complication and that other genetic or environmental factors are involved.

CRB1 genotype–phenotype correlation

Patients with classic RP, RP12, or LCA and *CRB1* mutations form a phenotypic continuum in which PPRPE and Coats-like exudative vasculopathy are inconsistent characteristics suggesting that other genetic and possibly environmental factors influence the expression of *CRB1* mutations. The absence of clear-cut null mutations on both *CRB1* alleles of patients with RP12 and classic arRP, and the presence of null mutations on both *CRB1* alleles in at least three LCA patients suggest that LCA may be associated with complete loss-of-function of CRB1, whereas RP12 and classic RP patients may have residual CRB1 function (Fig. 1B). Two LCA patients are homozygous for the most frequent mutation (C948Y) found in RP12 patients. In one patient with RP12 this mutation is found in combination with another missense mutation, and in two RP12 patients it is found together with a splice site mutation (IVS8+5G > A) that does not necessarily inactivate the mutant splice site completely (den Hollander et al 1999).

These findings suggest that C948Y is a severe mutation, leading to a severe phenotype when it is present homozygously. The C948Y mutation was also found homozygously in a patient with LCA in another study (Lotery et al 2001). C948Y changes the fourth conserved cysteine residue of the fourteenth EGF-like domain of CRB1, which is involved in the formation of disulfide bridges and thus in the correct folding of the EGF-like domain. Analogous to the ABCR model, we propose that a combination of severe *CRB1* mutations lead to LCA, a severe with a moderately severe *CRB1* mutation results in RP12, and combinations of severe/mild or of two moderately severe mutations may be associated with classic arRP (Fig. 1B). Additional mutation analysis and functional assays of individual mutations are required to test this hypothesis.

Conclusions

ABCA4 and *CRB1* are both mutated in patients with a spectrum of retinal dystrophies. The identification and functional analysis of *ABCA4* mutations has resulted in a genotype-phenotype correlation model which provides a valuable framework to elucidate the functional consequences of rare variants. This model, among clinical considerations, triggered us to search for *CRB1* mutations in RP12-related phenotypes as e.g. LCA and RP with Coats-like exudative vasculopathy. The routine screening of patients with autosomal recessive or isolated RP for mutations in *CRB1* might be important to reveal those patients that are at an increased risk to develop exudative retinal detachment. If this process is detected before it turns proliferative, cryotherapy can be used to prevent further progression.

Based on our results, we propose a CRB1 genotype-phenotype correlation model in which we hypothesize that combinations of severe with mild *CRB1* mutations might be associated with 'classic' arRP. For both ABCR and CRB1, small differences in activity seem to have dramatic effects on the clinical outcome, suggesting that small increments of protein activities in patients might have significant therapeutic effects.

Based on studies in mouse (Mata et al 2000) and *Drosophila* (Johnson et al 2002) deficient for *Abcr* or partially deficient for *Crumbs*, respectively, it has been proposed that patients with mutations in *ABCA4* or *CRB1* might be susceptible to light-induced photoreceptor degeneration and would therefore benefit from early molecular detection.

Acknowledgements

We thank Carolien Vink, Yvette J.M. de Kok, and Saskia D. van der Velde-Visser for expert technical assistance. We thank Rando Allikmets and Bernhard H.F. Weber for sharing unpublished data. These studies were made possible through grants from the Foundation Fighting Blindness Inc and the British Retinitis Pigmentosa Society.

References

Allikmets R 2000 Further evidence for an association of ABCR alleles with age-related macular degeneration. The International ABCR Screening Consortium. Am J Hum Genet 67: 487–491

Allikmets R, Hutchinson A, Jaakson K, Kuelm M, Pavel H 2001 Genotyping microarray (gene chip) for the *ABCR (ABCA4)* gene. Invest Ophthalmol Vis Sci 42:S714

Allikmets R, Singh N, Sun H et al 1997 A photoreceptor cell-specific ATP-binding transporter gene (*ABCR*) is mutated in recessive Stargardt macular dystrophy. Nat Genet 15:236–246

Briggs CE, Rucinski D, Rosenfeld PJ, Hirose T, Berson EL, Dryja TP 2001 Mutations in ABCR (ABCA4) in patients with Stargardt macular degeneration or cone-rod degeneration. Invest Ophthalmol Vis Sci 42:2229–2236

Cremers FPM, van de Pol DJR, van Driel M et al 1998 Autosomal recessive retinitis pigmentosa and cone-rod dystrophy caused by splice site mutations in the Stargardt's disease gene *ABCR*. Hum Mol Genet 7:355–362

Cremers FPM, van den Hurk JAJM, den Hollander AI 2002 Molecular genetics of Leber congenital amaurosis. Hum Mol Genet 11:1169–1176

den Hollander AI, ten Brink JB, de Kok YJM et al 1999 Mutations in a human homologue of *Drosophila* crumbs cause retinitis pigmentosa (RP12). Nat Genet 23:217–221

den Hollander AI, Heckenlively JR, van den Born LI et al 2001 Leber congenital amaurosis and retinitis pigmentosa with Coats-like exudative vasculopathy are associated with mutations in the crumbs homologue 1 (*CRB1*) gene. Am J Hum Genet 69:198–203

Fishman GA, Stone EM, Grover S, Derlacki DJ, Haines HL, Hockey RR 1999 Variation of clinical expression in patients with Stargardt dystrophy and sequence variations in the *ABCR* gene. Arch Ophthalmol 117:504–510

Fukui T, Yamamoto S, Nakano K et al 2002 ABCA4 gene mutations in Japanese patients with Stargardt disease and retinitis pigmentosa. Invest Ophthalmol Vis Sci 43: 2819–2824

Heckenlively JR 1982 Preserved para-arteriole retinal pigment epithelium (PPRPE) in retinitis pigmentosa. Brit J Ophthalmol 66:26–30

Izaddoost S, Nam S-C, Bhat MA, Bellen HJ, Choi K-W 2002 *Drosophila* crumbs is a positional cue in photoreceptor adherens junctions and rhabdomeres. Nature 416:178–182

Johnson K, Grawe F, Grzeschik N, Knust E 2002 Drosophila crumbs is required to inhibit light-induced photoreceptor degeneration. Curr Biol 12:1675–1680

Khan JA, Ide CH, Strickland MP 1988 Coats'-type retinitis pigmentosa. Surv Ophthalmol 32:317–332

Lewis RA, Shroyer NF, Singh N et al 1999 Genotype/phenotype analysis of a photoreceptor-specific ATP-binding cassette transporter gene, *ABCR*, in Stargardt disease. Am J Hum Genet 64:422–434

Lotery AJ, Jacobson SG, Fishman GA et al 2001 Mutations in the CRB1 gene cause Leber congenital amaurosis. Arch Ophthalmol 119:415–420

Martinez-Mir A, Paloma E, Allikmets R et al 1998 Retinitis pigmentosa caused by a homozygous mutation in the Stargardt disease gene *ABCR*. Nat Genet 18:11–12

Mata NL, Weng J, Travis GH 2000 Biosynthesis of a major lipofuscin fluorophore in mice and humans with *ABCR*-mediated retinal and macular degeneration. Proc Natl Acad Sci USA 97:7154–7159

Maugeri A, van Driel MA, van de Pol DJR et al 1999 The 2588G > C mutation in the *ABCR* gene is a mild frequent founder mutation in the western European population and allows the classification of *ABCR* mutations in patients with Stargardt disease. Am J Hum Genet 64:1024–1035

Maugeri A, Klevering BJ, Rohrschneider K et al 2000 Mutations in the *ABCA4* (*ABCR*) gene are the major cause of autosomal recessive cone-rod dystrophy. Am J Hum Genet 67:960–966

Maugeri A, Flothmann K, Hemmrich N et al 2002 The ABCA4 2588G > C Stargardt mutation: single origin and increasing frequency from South-West to North-East Europe. Eur J Hum Genet 10:197–203

Molday LL, Rabin AR, Molday RS 2000 ABCR expression in foveal cone photoreceptors and its role in Stargardt macular dystrophy. Nat Genet 25:257–258

Nasonkin I, Illing M, Koehler MR, Schmid M, Molday RS, Weber BHF 1998 Mapping of the rod photoreceptor ABC transporter (ABCR) to 1p21-p22.1 and identification of novel mutations in Stargardt's disease. Hum Genet 102:21–26

Paloma E, Martinez-Mir A, Vilageliu L, Gonzalez-Duarte R, Balcells S 2001 Spectrum of ABCA4 (ABCR) gene mutations in Spanish patients with autosomal recessive macular dystrophies. Hum Mutat 17:504–510

Papaioannou M, Ocaka L, Bessant D et al 2000 An analysis of ABCR mutations in British patients with recessive retinal dystrophies. Invest Ophthalmol Vis Sci 41:16–19

Pellikka M, Tanentzapf G, Pinto M et al 2002 Crumbs, the *Drosophila* homologue of human CRB1/RP12, is essential for photoreceptor morphogenesis. Nature 416:143–149

Rivera A, White K, Stohr H et al 2000 A comprehensive survey of sequence variation in the ABCA4 (ABCR) gene in Stargardt disease and age-related macular degeneration. Am J Hum Genet 67:800–813

Rozet J-M, Gerber S, Souied E et al 1998 Spectrum of ABCR gene mutations in autosomal recessive macular dystrophies. Eur J Hum Genet 6:291–295

Rozet J-M, Gerber S, Ghazi I et al 1999 Mutations of the retinal specific ATP binding transporter gene (ABCR) in a single family segregating both autosomal recessive retinitis pigmentosa RP19 and Stargardt disease: evidence of clinical heterogeneity at this locus. J Med Genet 36:447–451

Rudolph G, Kalpadakis P, Haritoglou C, Rivera A, Weber BH 2002 Mutations in the ABCA4 gene in a family with Stargardt's disease and retinitis pigmentosa (STGD1/RP19). Klin Monatsbl Augenheilkd 219:590–596

Shroyer NF, Lewis RA, Allikmets R et al 1999 The rod photoreceptor ATP-binding cassette transporter gene, ABCR, and retinal disease: from monogenic to multifactorial. Vision Res 39:2537–2544

Shroyer NF, Lewis RA, Yatsenko AN, Lupski JR 2001 Null missense ABCR (ABCA4) mutations in a family with Stargardt disease and retinitis pigmentosa. Invest Ophthalmol Vis Sci 42:2757–2761

Simonelli F, Testa F, de Crecchio G et al 2000 New ABCR mutations and clinical phenotype in Italian patients with Stargardt disease. Invest Ophthalmol Vis Sci 41:892–897

Sun H, Molday RS, Nathans J 1999 Retinal stimulates ATP hydrolysis by purified and reconstituted ABCR, the photoreceptor-specific ATP-binding cassette transporter responsible for Stargardt disease. J Biol Chem 274:8269–8281

Sun H, Smallwood PM, Nathans J 2000 Biochemical defects in ABCR protein variants associated with human retinopathies. Nat Genet 26:242–246

van den Born LI, van Soest S, van Schooneveld MJ, Riemslag FCC, de Jong PTVM, Bleeker-Wagemakers EM 1994 Autosomal recessive retinitis pigmentosa with preserved para-arteriolar retinal pigment epithelium. Am J Ophthalmol 118:430–439

Webster AR, Heon E, Lotery AJ et al 2001 An analysis of allelic variation in the ABCA4 gene. Invest Ophthalmol Vis Sci 42:1179–1189

Weng J, Mata NL, Azarian SM, Tzekov RT, Birch DG, Travis GH 1999 Insights into the function of Rim protein in photoreceptors and etiology of Stargardt's disease from the phenotype in *abcr* knockout mice. Cell 98:13–23

Yatsenko AN, Shroyer NF, Lewis RA, Lupski JR 2001 Late-onset Stargardt disease is associated with missense mutations that map outside known functional regions of ABCR (ABCA4). Hum Genet 108:346–355

DISCUSSION

LaVail: I have a question regarding the CRB1 localization. You mentioned that some people have found antibody artefacts in immunocytochemistry, and I'd agree that this is common. Do you see any other expression or localization there? If that is an artefact, where is the CRB1 actually localized?

Cremers: The artefact might be on the cone plasma membrane. We believe that CRB1 is expressed at the inner limiting membrane functioning as an anchor for cytoplasmic proteins and connecting photoreceptor cells with Müller cells. The cone has an envelope which contains among other proteins proteoglycans and similar molecules. This might be related to the non-specific reactions that other people are seeing with some antibodies to the same structure.

Gal: I have a more general question. I have the feeling that nowadays it is 'in' to call any RP gene an LCA gene. Nonetheless, by so doing we dilute this clinical diagnosis for, in the end, we include all recessive RP cases with early onset. Notably, in the papers first describing the two large RP12-famililes (van Soest et al 1994 for the Dutch family and Leutelt et al 1995 for the Pakistani family), the patients' phenotype was called autosomal recessive RP with para-arteriolar preservation of the RPE (PPRPE), and the diagnosis LCA was not considered for either of them. In the patients you studied, several had the unique phenotype of PPRPE. This is a hard criterion. On the other hand, you had some soft criteria, such as the early onset and severe course of the disease. Would it not be more sensible to classify these patients' disease according to the clearly unique phenotype (arRP with PPRPE) instead of the ill-defined age of onset and severity (LCA)?

Cremers: When we started a new project on genotype–phenotype analysis on LCA we looked at what had been published on these criteria. We have established a protocol to delineate the different groups. We took some of the classical criteria and grouped our patients. Our protocol is not fully accurate yet, but it would be a good idea to develop a widely accepted classification. In my own patient database I have a lot of patients that I can not classify well enough. In response to the comment of Dr Gal, patients with ARRP and PPRPE are not positioned in the LCA group, but patients with LCA, based on the fact that they had no vision at birth or lost vision before the age of 1 year, did show PPRPE.

Aguirre: You mentioned about limiting light exposure in these patients. Do you write letters to them recommending this, or to their physicians?

Cremers: Currently the patients themselves get general information on our molecular findings. As yet, we do not yet advise them specifically to avoid excessive light, but this might be important.

Aguirre: Many patients are now being advised for evaluation not to have the standard documentation with the six-field fundus photographs, but instead to have eight overlapping fields using the wide field camera. I have heard comments that this is more uncomfortable because the light is brighter, but they get a wider field of view. From what you are saying, this is a concern, because the light is much brighter than the light we are exposed to in the environment.

Bok: You showed a cartoon of CRB1 where it was depicted as a single-pass integral membrane protein with a very large extracellular mass. This suggests that it is either binding to the ECM or to itself. Do you have any gene disruption data yet to tell us about the cytoarchitecture of the retina?

Cremers: Studies of this kind are not performed in my group so I can't give details.

Bok: On the basis of the localization studies you described, presumably you don't think CRB1 is part of the zonula adherens, but is actually distal to that, in the plasma membrane of the inner segment.

Cremers: Yes, maybe just distal to the zonula adherens.

Bok: Do you predict that there would be some breakdown in the cytoarchitecture caused by a perturbation in cell–cell interactions?

Cremers: Yes. That is also seen in *Drosophila* CRB1 mutants, where in development, at a certain point the ommatidia elongate at a tremendous rate, with a 90° kink. In mosaic Crumbs mutant flies, these structures are not completely developed.

Bok: What about the rhabdomeres?

Cremers: It is the stalk region that is not growing well enough to produce this full structure.

Bok: You mentioned that homozygous CRB1 mutation is embryonic lethal in *Drosophila*.

Cremers: Yes, because it has a very important function in embryonic epithelia. Normally the cells are closely connected, but if CRB1 is missing their organization is disrupted.

McInnes: You called it a homologue, but from everything you have said it may be more accurate to say that it is the orthologue. Structural differences like the ones you showed are very common between flies and mammals, especially in the size of the protein.

Cremers: This might be true. However, in some regions CRB2 is more homologous to the *Drosophila* protein than CRB1 is and actually might be the true orthologue. At this point it is difficult to suggest which of the two is the actual orthologue.

Bolz: You showed some proteins that interact with CRB1. Have you looked at the expression pattern of the corresponding genes, or have you investigated them in patients?

Cremers: We have done some RT-PCR analysis. Many of the putatively interacting proteins show high but not exclusive expression in the retina.

Bolz: Is there any overlap of the loci with a known RP locus?

Cremers: Not as far as we know.

Daiger: I am interested in how you know that an amino acid substitution in the *ABCA4* gene is in fact pathogenic. Could you tell us about the background variation in this gene in a normal population? If you sequenced this gene in 100 individuals without disease, how many distinctly different amino acid substitutions would you find? How many of them are polymorphic in the classic sense that the allele frequency of the less frequent allele is 1% or greater? How many distinctly different protein haplotypes would be detected in that population of 200 chromosomes? Finally, in relation to this, what is your definition of a mutation in this context?

Cremers: A Japanese group recently analysed the *ABCA4* gene in normal individuals (Iida et al 2002). Also, data are known for a German control population which was tested for the presence of known *ABCA4* mutations. Pathologic mutations are found in ∼ 5% of healthy individuals.

Daiger: How do you define 'mutations' and 'polymorphisms' in this context?

Cremers: About 50 mutations have been tested in a rather crude *in vitro* functional assay by looking at whether the protein is expressed and is transported to the cell membrane, and whether there is ATPase activity. In these tests most of these mutations have shown deficiencies in the protein.

Daiger: Isn't it true that some of the things that you are calling 'polymorphisms' in the laboratory also show activity diminution? What leads you to say whether or not they are pathogenic?

Cremers: The 2588G > C allele I described always is found together with another mutation. Both show a deficiency in this artificial test. In humans with Stargardt disease we find the 2588G > C allele alone in a few cases without the what we think is a polymorphic variant, but not vice versa. From this interpretation it is almost certain that it is the 2588G > C mild allele that is pathological.

Daiger: So you use the term 'polymorphic' to describe a variant that is non-pathogenic, independent of its frequency, and you use 'mutation' for a variant that is pathogenic, again independent of its frequency.

Cremers: Yes.

McInnes: I don't think we should reinvent genetic terminology in this way. A polymorphism is any allele found in 1% or more of the population.

Daiger: I am not trying to hone in on the accuracy of nomenclature. Nomenclature in this field has changed radically in the last five years in a way that has horrified population geneticists. But what is important is that we are clear about what we are talking about. How many amino acid substitutions are there in Caucasian populations which by any criteria are pathogenic, and how many are non-pathogenic? This is what has worried me most about the field of ABCA4: the difficulty of defining what is truly a pathogenic amino acid substitution.

Swaroop: I think we should refrain from saying anything is a mutation unless we know it is pathological. In the absence of this evidence of a causative change we use the term 'sequence variant'.

Cremers: I agree. But what about the IVS38-10T>C polymorphism that I indicated? If this is found in 25 alleles of 500 Stargardt disease patients and is not found in 500 controls, what more evidence do we need that this variant is in linkage disequilibrium with the actual (unknown) pathologic DNA change? To not use this as a diagnostic marker would be unrealistic.

McInnes: We have to remain orthodox in our language. A mutation is any change in a DNA sequence: it has nothing to do with pathogenicity although we all tend to use it that way. The trouble is that the loose terminology will lead to loose thinking.

Bird: As a clinician it is difficult listening to others' nomenclature. It is generally believed in ophthalmology that Stargardt disease is found in about 1 in 6500 of the population. This means a carrier rate of about one in 40. This would fit, in general, with the risk to the child of an affected individual being about 1 in 80, which fits with our experience. But I then find it difficult when the implication is that there is a disease-causing mutation that is much more common than 2.5% of the population. If the carrier rate is 1 in 40, and a sequence change that is thought to be disease causing is more common than that, I have a hard time understanding it, if you assume one mutation per allele.

Daiger: The cystic fibrosis allele frequency among Caucasians is 3–5%, and the sickle cell allele frequency is 5–7% among African Americans. It is not unheard of for recessive diseases to have carrier frequencies in the 5–10% range.

Cremers: Most of the mutations are mild alleles, otherwise there would be many more Stargardt cases. What needs to be done with this chip is to test a few thousand of these cases and also a couple of thousand controls to see what the relationship is. This might also resolve the question of whether *ABCA4* mutations play a role in AMD.

Bhattacharya: As long as the controls have been tested very thoroughly for any signs of AMD.

Daiger: I can't resist adding that one of the most difficult statistical problems is comparing a common disease such as AMD with a gene that has common variants in it. With regards to the cost of analysis, I don't think we disagree much on the numbers. If you limit the study of autosomal dominant RP to 'common' mutations, and define 50 or so of those, you could reduce these to the gene chip technology. The use of the chip may be only $75, but this doesn't include preparation of DNA, reports, personnel costs and so on.

Cremers: These costs have already been paid for. The laboratory and staff are already there.

Daiger: When you give a quote of $75 to test 380 variants, are you including overhead and personnel costs in that?

Cremers: Yes, the Estonian company has done this for this sum which is a research price. In most situations we already have the diagnostic set-up. There is a parallel group that does the diagnostics more-or-less independently from my group and they already have the infrastructure for this analysis.

Daiger: My point is that someone had to pay for that in the first place.

Zack: I have a more general comment in terms of the regulatory issues of CLIA certification. We have had problems here in Baltimore with university regulations that we can't tell the patient anything unless the results are coming from a CLIA-certified lab. This means that if we have a result we can't tell them. When a patient wants something tested, how do you deal with this?

Daiger: There is a practical answer. In addition to getting your own lab CLIA certified, which isn't as difficult as it sounds, there are certain reference laboratories that are CLIA certified. If you find a mutation, the patient then sends this lab $500 and a blood sample and they will confirm the existence of that mutation. This may not be completely satisfactory, but it is one way of doing this. You can tell the lab where to look for the mutation.

Cremers: This is the situation we are faced with. We have a DNA diagnostic division. They take the blood sample, we obtain half the DNA sample to do a complex mutation analysis and then we give the result back to them. Alternatively, if the blood sample was sent directly to us, another blood sample is drawn and directly sent to the DNA diagnostics facility. They repeat the analysis and send out the result.

Daiger: Ted Dryja, how do you handle patients?

Dryja: I am nervous about this issue because we don't have a general rule. I feel uncomfortable informing patients of results from our research lab because it is not operated with the quality controls that are required for a clinical testing lab. I estimate that we probably have an error rate of 1%, but many patients will take the results from a research lab and assume they are absolutely correct. What is our legal liability here? In reality, most of the time we don't inform the patients. They may have given blood 5–10 years ago; we have samples from thousands of patients; it would take an enormous amount of effort to just mail out results to patients each year.

Daiger: We have official approval from our Human Subjects Committee to initiate a process to go back to the individuals and inform them that we have specific genotype information on them relevant to their retinal disease. We ask them whether they want this information released to a third party, who has to be a 'knowledgeable healthcare professional'. With that permission from the patient, one of our standard forms goes out to the clinician. It is cumbersome, but our human subjects committee has been comfortable with this for the last few years.

References

Iida A, Saito S, Sekine A et al 2002 Catalog of 605 single-nucleotide polymorphisms (SNPs) among 13 genes encoding human ATP-binding cassette transporters: ABCA4, ABCA7, ABCA8, ABCD1, ABCD3, ABCD4, ABCE1, ABCF1, ABCG1, ABCG2, ABCG4, ABCG5, and ABCG8. J Hum Genet 47:285–310

Leutelt J, Oehlmann R, Younus F et al 1995 Autosomal recessive retinitis pigmentosa locus maps on chromosome 1q in a large consanguineous family from Pakistan. Clin Genet 47:122–124

van Soest S, Ingeborgh van den Born L, Gal A et al 1994 Assignment of a gene for autosomal recessive retinitis pigmentosa (RP12) to chromosome 1q31-q32.1 in an inbred and genetically heterogeneous disease population. Genomics 22:499–504

What should a clinician know to be prepared for the advent of treatment of retinal dystrophies?

Alan C. Bird

Department of Clinical Ophthalmology, Moorfields Eye Hospital, City Road, London EC1V 2PD, UK

Abstract. It is now evident that several forms of therapy have influenced inherited retinal degeneration in animals. These are gene therapy, cell transplantation, slowing of cell death using growth factors and pharmacological approaches. There are firm proposals to initiate gene therapy in the foreseeable future. For therapy to be successful a variety of attributes of disease must be established so that the full benefits of research can be realised in clinical practice. These can be considered as identification of the causative genes, knowledge of the disease mechanisms and detection of the therapeutic effect.

2004 Retinal dystrophies: functional genomics to gene therapy. Wiley, Chichester (Novartis Foundation Symposium 255) p 85–94

It is now evident that several forms of therapy have influenced inherited retinal degeneration in animals. These are gene therapy, cell transplantation, slowing of cell death using growth factors and pharmacological approaches (Faktorovich et al 1990, LaVail et al 1998, Green et al 2001, Ali et al 2000, Acland et al 2001, Kwan et al 1999, Lund et al 2001, Frasson et al 1999). There are firm proposals to initiate gene therapy in the foreseeable future. For therapy to be successful a variety of attributes of disease must be established so that the full benefits of research can be realised in clinical practice. These can be considered as identification of the causative genes, knowledge of the disease mechanisms and detection of the therapeutic effect.

Genes responsible for disease

Many genes are now known to cause retinal dystrophies. To document the mutant genes responsible for retinal dystrophies in any population is necessary for a genotyping service linked to a genetic register. Genotyping is not currently available, and the funding of this activity may be difficult. Currently it should be possible to identify the responsible gene in about 50% of families with autosomal dominant disease, and most if not all X-linked disease. For autosomal recessive disease the number is uncertain. In the commonest macular disease, Stargardt macular dystrophy, for which the gene is known, the proportion of mutations detected is low. This is in part because the gene is large and highly polymorphic such that there is uncertainty as to the significance of base changes to disease. It is clearly desirable for the mutation to be known in any patient entering a treatment trial, and essential for gene therapy.

Knowledge of the disease mechanisms

The following should be established:

- the cell expressing the mutant gene and the target cell of disease
- whether the gene causes gain or loss of function and the biochemical mechanisms of disease
- the temporal profile of functional loss and cell death

The cell expressing the mutant gene and the target cell of disease

In most disorders the cell harbouring the mutation is recognizably the same as the cell that is most affected by the disease. However this should not be assumed. In Stargardt disease the mutant gene is expressed in the photoreceptor cells, and yet the cell that initiates the process of visual loss is the retinal pigment epithelial cell that becomes overloaded with autofluorescent material. This can be detected by autofluorescence imaging, and is the first sign of disease.

Whether the gene causes gain or loss of function

Whether disease is due to haploinsufficiency, or gain of function will determine a suitable approach to gene therapy. Knowledge of this will come from work by molecular geneticists, cell biologists and biochemists.

In haploinsufficiency the objective is to insert a gene that will achieve long-term and appropriate expression of a protein that is correctly targeted. There is ample evidence that this can be achieved in animal models with subretinal injections and long-term rescue of function has been achieved.

Many instances of gain of function have been described. This has been nicely illustrated with mutant RetGC1 and GCAP1 (Newbold et al 2001, Wilkie et al 2000). In such cases the objective would be to block gene expression using ribozymes by which mRNA from the mutant gene would be cleaved with consequent reduction of the relevant mutant protein. Different ribozymes have variable specificity and efficiency, but observations on rodent models indicate that this is an approach with promise (LaVail et al 2000). In general it has been assumed that loss of function is a feature of recessive disease, and gain of function of dominant disease. However this is not always the case. In retinitis pigmentosa due to mutations in RDS (Goldberg et al 1998), and PRPF31, it appears that the degeneration is due to loss of function.

Recently it has been shown that some dogs with X-linked PRA, due to mutations in ORF15 of the RPGR gene, have an expanded rough endoplasmic reticulum (RER); a situation homologous to liver in α chymotrypsin deficiency (Zhang et al 2002). In this situation, it is possible that there is a combination of mechanisms, one of which may be more critical than the other.

Temporal profile of cell death

It is evident that gene therapy will only succeed if the target cell of disease is physically present. It has been shown that in some disorders loss of function is due to cell dysfunction at least at some stage in the disease process, whereas in others it signals cell death (Massof & Finkelstein 1981, Lyness et al 1985). The two forms of disease were shown in dominant RP. In the first was termed as type 1 or diffuse, and the second regional or type 2. In the first there was widespread loss of scotopic function but relative preservation of photopic sensitivities. In the second there was patchy and equal loss of rod and cone mediated sensitivities. The rhodopsin levels were assessed by relfectometry (Kemp et al 1988). In type 2 loss of rod sensitivities, rhodopsin level was proportional to the loss of function. By contrast, in type 1 rhodopsin levels were much higher than would have been predicted from functional measurements; patients with more than three log units elevation of threshold had near normal concentrations of bleachable rhodopsin. In type 2 loss of function was determined by light catch, and all other functions of the visual system were normal. In type 1 loss of function might be due to loss of transduction gain or noisy photoreceptors due possible to constitutive activity of rhodopsin. The form of the disorder was consistent within a family implying that the differences were intrinsic to the disease.

The implications for therapy are self-evident. In type 1, gene therapy might cause recovery of function, whereas in type 2 it would serve only to slow down the kinetics of loss. In type 1, cell transplantation would be inappropriate, but would be suitable for type 2.

Thus the relationship between cell death and functional loss must be established for each disorder for which treatment is contemplated. With respect to photoreceptor cells this can be achieved with autofluorescence imaging, or optical devices that allow visualisation of the outer retina (von Rückmann et al 1995, Fitzke 2000). The latter might be achieved using confocal optical coherence tomography or corrective optics. It is believed that acquisition of autofluorescence in the retinal pigment epithelium (RPE) is determined by its metabolic activity, which is dependent largely upon photoreceptor outer segment renewal. Thus loss of autofluorescence implies photoreceptor cell death or at least outer segment loss. The presence of normal autofluorescence implies that the photoreceptor cell layer is intact and that there are outer segments that are being constantly renewed. Abnormally high levels of autofluorescence indicate inability of the RPE to process phagosomal material. That function can be poor in patients with an apparently intact outer retina is illustrated by normal autofluorescence in a 15 year old with Leber amaurosis whose vision has not been recorded at better than light perception from early life.

In most outer retinal dystrophies the RPE remains physically present throughout disease, despite appearing very abnormal following loss of photoreceptor cells. Exceptions to this would include choroideremia.

Detection of the therapeutic effect

Under certain circumstances gene therapy may cause gain of function. In other disorders there may be reversal of a specific clinical attribute of disease such as the photophobia so characteristic of some with mutations in RetGC1. Both of these should be readily evident clinically. By contrast the effect may be slowing or cessation of progression. This would be much more difficult to detect. It will require very accurate and reproducible tests of function and recording of the physical state of the retina. It would also be required that the tests be well tolerated by patients. Many psychophysical tests of function require a great deal from the patient, whereas electrophysiology, being objective is somewhat easier. Some laboratories have good longitudinal recording of function, and patients could be selected for therapy in whom the time-course of the disease is known, and reliability of recording verified.

Internationally agreed protocols for tests of electrophysiological responses have been established (Marmor & Zrenner 1993) but for psychophysical attributes have not; this clearly needs to be addressed. Methods by which alteration of profiles of cell loss have only recently been addressed.

Clinical setting for treatment

It is evident that a large well-documented patient pool is necessary to realise the full benefits of treatment, and should thus be introduced into clinical practice. It will be required that mutations be known and that the patients have well-documented disease. Techniques for recording the treatment effect such as electroretinography, psychophysics and specialized imaging must be available. This requires extensive registers of disease with all the available data recorded and easily accessible by the clinicians involved. The testing equipment exists although not all major centres have access to all equipment needed. The techniques are practiced by many centres; reproducibility and progression have been well documented. Finally the centres should be easily accessed by patients.

To achieve these objectives it will take a great deal of time and effort on the part of clinicians, scientists and patients. This activity should take place in parallel with the establishment of clinical services for inherited ocular diseases.

Conclusions

It is evident that biological approaches to treatment have achieved success in experimental studies. There is also reason to hope that a minor change in the metabolic environment of the retina can have a profound and long lasting effect on the course of disease. The extension of work from rodents to larger animal models is in its early stages, in part because identification and generation of such models is very recent. This extension is another important precedent to human trials since larger eyes provide the opportunity to develop delivery systems and assess safety.

Each therapeutic approach has its potential advantages and disadvantages. As the work evolves it is possible that all three techniques will be used in combination. For example, transplantation of transgenic cells that express a growth factor over long periods may prove viable.

Although the biological treatment may be a future dream, it must be encouraging to the clinician and their patients that progress is being made in providing therapy for these intractable diseases. It is important that the clinical community is in a position to take advantage of these advances when clinical application can be justified. It will require a large number of patients with well-characterized disease, protocols by which the benefits of treatment can be tested and centres that are capable of carrying out such tasks. The model is illustrated by the therapeutic trial of vitamin A, which was a monumental task that required immense discipline (Berson et al 1993).

References

Acland GM, Aguirre GD, Ray J et al 2001 Gene therapy restores vision in a canine model of childhood blindness. Nat Genet 28:92–95

Ali RR, Sarra GM, Stephens C et al 2000 Restoration of photoreceptor ultrastructure and function in retinal degeneration slow mice by gene therapy. Nat Genet 25:306–310

Berson EL, Rosner B, Sandberg MA et al 1993 A randomized trial of vitamin A and vitamin E supplementation for retinitis pigmentosa. Arch Ophthalmol 111:761–772

Faktorovich EG, Steinberg RH, Yasumura D, Matthes MT, LaVail MM 1990 Photoreceptor degeneration in inherited retinal dystrophy delayed by basic fibroblast growth factor. Nature 347:83–86

Fitzke FW 2000 Imaging the optic nerve and ganglion cell layer. Eye 14:450–453

Frasson M, Sahel JA, Fabre M, Simonutti M, Dreyfus H, Picaud S 1999 Retinitis pigmentosa: rod photoreceptor rescue by a calcium-channel blocker in the rd mouse. Nat Med 5:1183–1187

Goldberg AF, Loewen CJ, Molday RS 1998 Cysteine residues of photoreceptor peripherin/rds: role in subunit assembly and autosomal dominant retinitis pigmentosa. Biochemistry 37: 680–685

Green ES, Rendahl KG, Zhou S et al 2001 Two animal models of retinal degeneration are rescued by recombinant adeno-associated virus-mediated production of FGF-5 and FGF-18. Mol Ther 3:507–515

Kemp CM, Jacobson SG, Faulkner DJ 1988 Two types of visual dysfunction in autosomal dominant retinitis pigmentosa. Invest Ophthalmol Vis Sci 29:1235–1241

Kwan ASL, Wang S, Lund RD 1999 Photoreceptor layer reconstruction in rodent model of retinal degeneration. Exp Neurol 159:21–33

LaVail MM, Yasumura D, Matthes MT 2000 Ribozyme rescue of photoreceptor cells in P23H transgenic rats: long-term survival and late-stage therapy. Proc Natl Acad Sci USA 97:11488–11493

LaVail MM, Yasumura D, Matthes MT et al 1998 Protection of mouse photoreceptors by survival factors in retinal degenerations. Invest Ophthalmol Vis Sci 39:592–602

Lund RD, Adamson P, Sauve Y 2001 Subretinal transplantation of genetically modified human cell lines attenuates loss of visual function in dystrophic rats. Proc Natl Acad Sci USA 98:9942–9947

Lyness AL, Ernst W, Quinlan MP et al 1985 A clinical, psychophysical and electroretinographic survey of patients with autosomal dominant retinitis pigmentosa. Br J Ophthalmol 69: 326–339

Marmor MF, Zrenner E 1993 Standard for clinical electro-oculography. International Society for Clinical Electrophysiology of Vision. Doc Ophthalmol 85:115–124

Massof RW, Finkelstein 1981 Two forms of autosomal dominant primary retinitis pigmentosa. Doc Ophthalmol 51:289–346

Newbold RJ, Deery EC, Walker CE et al 2001 The destabilization of human GCAP1 by a proline to leucine mutation might cause cone-rod dystrophy. Hum Mol Genet 10:47–54

von Rückmann A, Fitzke FW, Bird AC 1995 Distribution of fundus autofluorescence with a scanning laser ophthalmoscope. Br J Ophthalmol 79:407–412

Wilkie SE, Newbold RJ, Deery E et al 2000 Functional characterization of missense mutations at codon 838 in retinal guanylate cyclase correlates with disease severity in patients with autosomal dominant cone-rod dystrophy. Hum Mol Genet 9:3065–3073

Zhang Q, Acland GM, Wu WX et al 2002 Different RPGR exon ORF15 mutations in Canids provide insights into photoreceptor cell degeneration. Hum Mol Genet 11:993–1003

DISCUSSION

McInnes: I think your model of ROM1 and RDS is in essence correct. Our initial antibody didn't pick it up in cones, but Bob Molday has shown it is there. However, in the knockout mouse, the cones are fine even though the rods are disappearing. If you are keeping the cells from dying in your type 2 patchy loss of function, why is this merely going to delay loss? If you correct the defect, then presumably they ought not to die.

Bird: In type 2 in which loss of function is due to cell death, gene therapy will slow down or delay visual loss, as opposed to type 1 in which the cells are dysfunctional but present you could induce recovery of function. This is important to the detection of the therapeutic effect. Obviously, what we would love to do is to cause recovery of function. This would be very easy to measure. If we are establishing priorities we want to choose the disorders where there are large populations of cells that aren't working first. I am not saying that we shouldn't do gene therapy in type 2, but the best we can expect there is delay of functional loss, and this will be much harder to measure and to be certain of.

Aguirre: In your population of autosomal dominant RP, I'd imagine that many of these patients have rhodopsin mutations. What is the percentage of patients that have type 2 versus type 1?

Bird: The work was done back in the 1980s. Rhodopsin mutations were found in several of those families, and we found that the mutations were in both categories. Most were type 1.

Daiger: I agree with the need for the patient registries that you described. Unfortunately, this would be difficult to do in the USA, where we have human subjects constraints and a new law called HIPA, with draconian consequences for loss of confidentiality of patient material. Will this be easier in Europe?

Bird: We are extraordinarily fortunate in London in that Marcel Jay established a register in the 1970s, which now contains over 5000 families. Some of these are very big, with 100–200 affected, traced back to the mid-19th century. We have close to 100% ascertainment of X-linked RP on our register. We have not seen an X-linked RP that we cannot connect to one of our families for about five years. We have about 60% of dominant disease on our register. There was a law proposed in Europe to make genetic registers illegal, so we were just going to call it a diagnostic index. Fortunately, this law was rejected. There is a data protection act, but we are allowed to store this information. We can be consulted just as a doctor consults another doctor, and we are allowed to release information that is medically relevant.

Daiger: It is an evolving issue in the USA, but it looks as if in order to have a name in such a registry with any identifying information, we need written consent from those individuals.

Bird: All of our subjects now must give written consent and can request at any time that their name be removed. In Europe the desire of insurance companies to have access to these lists will be resisted very strongly.

Hauswirth: Regarding your comment on the gene therapy application in type 2 only being able to slow down cell death, I agree this is the current way of thinking. But there are two scenarios in which therapy might be more applicable to type 2. The first one would be if it became possible to diagnose type 2 very early in life: then you might be able to deal with the toxic product and not have the disease at all. The second would involve gene correction therapy. This is where a mutant allele that is making a dominant negative product is turned into a normal gene. This should stop the disease in its tracks.

Bird: In type 2, the loss of rhodopsin is the only factor determining loss of sensitivity. Everything else appears to be normal. It would certainly be possible to stop the disease progressing. This work all dates from the early 1980s and has attracted little attention since.

Bok: Knowing your level of curiosity, I am surprised that you haven't gone back and correlated the two types with the genotypes, which you must have in your files.

Bird: The mutations known were limited to rhodopsin. It was a small sample of about 12 families, and it was a huge amount of work. We found rhodopsin mutations in both categories.

Dryja: Do different rhodopsin mutations cause type 1 and type 2?

Bird: Yes.

Dryja: And no one mutation can be in both groups?

Bird: That's correct. It is consistent within families.

Bhattacharya: Did he ever study any of the 19q families?

Bird: No, that is precisely what we are doing now: we are recruiting the 19q families for study.

Zack: Looking to the future rather than to the past, it seems that there is a nice distinction between type 1 and type 2. Is there a technology that could be developed for physicians to be able to make this distinction more accurately and easily?

Bird: One diagnostic criterion is the temporal profile of visual loss as a standard question. If someone says they had no problems by day until they were in their 30s but they have never really seen very well at night, this would be an indication that it is probably type 1. I recognize that patients are not always very good witnesses, and this is indefinite information. If you see someone who has had poor night vision for 30 years and there is not much pigmentation, it is probably type 1. What we need is some easy, rapid way of distinguishing one from another. Of the techniques available to us now, I would argue that autofluorescence looks like the easiest way of doing this. OCT certainly is available, but autofluorescence is quicker and easier.

Sieving: The difference between type 1 and type 2 is interesting, but one of the features on your graph of type 2 is that sensitivity loss correlates with loss of quantum catch, and in type 1 it does not. However, type 1 cluster together at the top of the graph, in a region that would correspond with cone sensitivity. This indicates that there is no gradient of loss in type 1 cases and implies that any degree of constitutive activity leads to extensive rod desensitization. This would make it difficult to restore rod function unless the deficit is totally reversed. By contrast, in RPE65 gene transfer the strategy is to provide something that is otherwise absent (i.e. RPE65 activity), and one would expect a gradient of restoration of function which would not require total reversal of the condition. I think the restoration of vision in type 1 would be a difficult target; half-restoration would not suffice; you'd have to completely restore it.

Dryja: I had the impression that the measurements of rhodopsin in the retinas of patients tended to show that the rod photoreceptor cells are present in those cases. There is plenty of rhodopsin but the rods aren't working. The hope is that we might be able to regain rod function.

Bird: If, for example, it was a noisy cell because rhodopsin was constitutively active, and you could suppress the expression of the noisy rhodopsin, this might work.

Sieving: Yes, but you would have to completely suppress the expression of noisy rhodopsin.

Dryja: In the case of a noisy cell it would probably be very difficult. But if you provided functioning pigment to patients who have a non-functioning pigment, it would be theoretically beneficial.

Bird: Certainly, the rhodopsin is bleaching and regenerating since the reflectometric assessment is dependent on this.

Aguirre: You mentioned that fundus autofluorescence would be a good way of assessing patients with type 1 disease. Why? If you have rhodopsin, you would expect that the shedding of photoreceptor tips is ongoing even if it is not functioning, so there would be a normal level of fundus autofluorescence.

Bird: Autofluoresence can be used to measure physical loss of cells. Take Stargardt disease as an example. If you say you have a treatment for it, if an area of heightened autofluorescence disappears without there being evidence of cell death, you have done something to that patient that would never occur without treatment. A possible measure of therapeutic effect is disappearance of heightened autofluorescence without accompanying cell death. This would be the same for age-related macular degeneration (AMD). Focal increases in autofluoresence always leads to geographic atrophy in the natural history of disease.

Bhattacharya: Would you recommend that in all types of retinal disease that fluorescence imaging should be a standard test?

Bird: Yes. Any standard phenotyping needs to be quick and easy for patients. The initial measurements of rhodopsin levels done by Harris Ripps demanded a huge amount of the patient. A great deal of psychophysical techniques demands a great deal from the patient. It is a matter of choosing protocols that are acceptable to patients, and autofluorescence imaging and optical coherence tomography (OCT) are both straightforward for the patient. I would argue that they should be part of the routine of phenotyping patients.

Bhattacharya: Is this a standard in the USA?

Sieving: Both autofluorescence imaging and OCT appear to be quite useful approaches, but unfortunately they are not yet used frequently in the USA.

Farber: How much does the imaging equipment cost?

Bird: It depends how you negotiate with the company. We reckon we can get it for £50 000, although the shelf price is greater than this.

Gal: In the case of rhodopsin mutations in autosomal dominant RP, most likely all rod photoreceptors express the mutant protein, but only some of them die, occasionally in a regional pattern. Could you speculate about how much is genetically and non-genetically determined in this process?

Bird: You are talking about the distribution of cell loss. I think this is a fundamental question in all retinal dystrophies. Why does RP usually start at 12 degrees of eccentricity? Why should some people have a physically separate ring of atrophy around the fovea at three degrees? What is peculiar about those two loci within the retina? These are features of RP that have been well documented. The physical distribution of cell death is strange and distinctive, and there is very little information as to why that should be. Is a rod at 12 degrees fundamentally different from one at 18 degrees? It was interesting looking at the rhodopsin expression as they truncated the promoter region in which different distributions of expression occurred.

Sieving: We have all thought about the issue of divergent phenotypes for the same gene. For us this came up most recently in the context of the family that we published in *Genomics* recently with a mutation in ORF15 of the RPGR gene, in which there was a consistent phenotype of macular atrophy for this XL-RP3 gene. (Ayyagari et al 2002). This macular degeneration phenotype is remarkably divergent from the global retinal degeneration that is otherwise the hallmark of X-linked RP3 disease. The earliest stages showed incipient atrophy beginning just para-foveal, at a time when the acuity was still 20:20, and it progressed to circumscribed macular atrophy with minimal peripheral retinal involvement. One might deduce from that the topographical mapping of macular degeneration requires both rods and cones to be present for the earliest insult to occur, and that there is an important dynamic involved between these two cell types and the RPE.

Thompson: Are any of these families with macular degeneration big enough that you could consider mapping a second locus in the family, based on the assumption that the mutation in the primary gene is causing the degeneration, and that a polymorphism in a modifier gene is being revealed by the primary mutation as well?

Bird: That is relatively straightforward if there is bimodal expression disease: that is, it is either very severe or it is very mild, as occurs with 19q. On the other hand, in most of the families we look at it appears to be a Gaussian curve of severity. So you can't say we have to compare these two categories of affection in the family to see how they are different.

Reference

Ayyagari R, Demirci FY, Liu J et al 2002 X-linked recessive atrophic macular degeneration from RPGR mutation. Genomics 80:166–171

Role of subunit assembly in autosomal dominant retinitis pigmentosa linked to mutations in peripherin 2

Robert S. Molday, Laurie L. Molday and Christopher J. R. Loewen

Department of Biochemistry and Molecular Biology and Department of Ophthalmology and Visual Sciences, University of British Columbia, 2146 Health Sciences Mall, Vancouver, BC V6T 1Z3, Canada

Abstract. Peripherin 2 is a photoreceptor-specific membrane protein implicated in outer segment disk morphogenesis and linked to various retinopathies including autosomal dominant retinitis pigmentosa (ADRP). Peripherin 2 and ROM1 assemble as a mixture of core noncovalent homomeric and heteromeric tetramers that further link together through disulfide bonds to form higher order oligomers. These complexes are critical for disk rim formation and outer segment structure through interaction with the cGMP-gated channel and other photoreceptor proteins. We have examined the role of subunit assembly in peripherin 2 targeting to disks, outer segment structure, and photoreceptor degeneration by examining molecular and cellular properties of peripherin 2 mutants in COS-1 cells and transgenic *Xenopus laevis* rod photoreceptors. Wild-type (WT) and the ADRP-linked P216L mutant were transported and incorporated into newly formed outer segment disks of transgenic *X. laevis*. The P216L mutant, however, induced progressive outer segment instability and photoreceptor degeneration possibly through the introduction of a new N-linked oligosaccharide chain. In contrast, the C214S and L185P disease-linked, tetramerization-defective mutants, were retained in the inner segment, but did not affect outer segment structure or induce photoreceptor degeneration. Together, these results indicate that peripherin 2 mutations can cause ADRP either through a deficiency in WT peripherin 2 (C214S, L185P) or by a dominant negative effect on disk stability (P216L).

2004 Retinal dystrophies: functional genomics to gene therapy. Wiley, Chichester (Novartis Foundation Symposium 255) p 95–116

Peripherin 2, also known as peripherin/rds, is a photoreceptor specific membrane protein localized along the rims and incisures of rod and cone outer segment disks (Arikawa et al 1992, Molday et al 1987). It plays a central role in outer segment disk morphogenesis and photoreceptor viability since *rds* mice homozygous for the disrupted peripherin 2 gene fail to produce outer segments and heterozygous *rds* mice with decreased levels of peripherin 2 have shortened, highly disorganized outer segments consisting of whorls of disk membranes (Connell et al 1991, Hawkins et al 1984, Sanyal & Jansen 1981, Travis et al 1989). The absence or disorganization of outer segments leads to progressive photoreceptor degeneration.

The importance of peripherin 2 in photoreceptor viability is further highlighted by the finding that over 40 different missense, nonsense, insertion, deletion and splice-site mutations in the human peripherin 2 gene (*ROS*) have been linked to a variety of clinically defined retinopathies (Farrar et al 1991, Kajiwara et al 1991, Saga et al 1993, Weleber et al 1993). Mutations in peripherin 2 account for up to 5% of the cases of autosomal dominant retinitis pigmentosa (ADRP). In addition, selected mutations in peripherin 2 have been associated with clinically distinct pattern dystrophies including macular dystrophy, bull's eye maculopathy, butterfly-shaped dystrophy, cone-rod dystrophy, adult vitelliform macular dystrophy, pigment dystrophy, autosomal retinitis punctata albescens and others. Many of these retinal dystrophies appear to be phenotypic variations of the same disease (Weleber et al 1993). In most cases, disease-linked mutations in peripherin 2 lead to progressive loss of both rod and cone photoreceptors. However, in several cases, one cell type is affected more severely than the other, suggesting that there are differences in the functional properties and protein interactions of peripherin 2 in rod and cone photoreceptors (Cheng et al 1997, Cideciyan et al 2000, Jacobson et al 1996).

Photoreceptor outer segments also contain a structurally related protein known as ROM1 which interacts with peripherin 2 to form a heteromeric complex at the rims of disk membranes (Bascom et al 1992, Goldberg et al 1995, Moritz & Molday 1996). Although initial studies suggested that ROM1 is only expressed in rod photoreceptors (Bascom et al 1992), subsequent immunocytochemical studies indicate that ROM1, like peripherin 2, is present in both cone and rod cells (Moritz & Molday 1996). Mutations in the *ROM1* gene have been linked to a digenic form of ADRP (Kajiwara et al 1994, Dryja et al 1997). Individuals who inherit a *ROM1* null allele or a G113E missense mutation together with a L185P peripherin 2 mutation are afflicted with a severe form of ADRP, whereas individuals who inherit either the *ROM1* or peripherin 2 mutation are essentially normal. To date, mutations in *ROM1*, alone, have not been conclusively shown to cause a monogenic retinal disease.

In order to understand the role of peripherin 2 and ROM1 in photoreceptor outer segment structure, function and renewal, and elucidate the molecular mechanisms underlying various diseases linked to mutations in peripherin 2 and ROM1, we and other research groups have examined the molecular and cellular properties of these membrane proteins. Here, we review some recent findings on the structure of the peripherin 2/ROM1 complexes and their role in outer segment structure and morphogenesis. We also discuss molecular and cellular mechanisms by which selected mutations in peripherin 2 cause ADRP.

Structural features of peripherin 2 and ROM1 subunits

Peripherin 2 and ROM1 are members of the widely distributed tetraspanin family of membrane proteins that have been implicated in such diverse cellular functions as cell development, fertilization, proliferation, signalling, mobility, adhesion and receptors for pathogens including viruses (Hemler 2001). Tetraspanins are characterized by the presence of several conserved structural features including the N and C terminal segments exposed on the cytoplasmic surface of cell membranes, four transmembrane segments (M1–M4), two exocytoplasmic (extracellular or lumen) segments — a small segment (EC-1) between M1 and M2 and a large domain (EC-2) between M3 and M4, and four invariant cysteine residues in the EC-2 domain, two of which are part of a conserved CCG motif (Hemler 2001, Seigneuret et al 2001).

The high resolution structure of the EC-2 domain of the tetraspanin protein CD81, a receptor for hepatitis C virus has been determined by X-ray crystallography (Kitadokoro et al 2001). It consists of five α helices arranged in a stalk and head configuration. The stalk subdomain is involved in noncovalent interactions responsible for homodimer formation, whereas residues in the head subdomain function in the binding of the hepatitis C virus.

The current topological model for peripherin 2 is shown in Fig. 1. Peripherin 2 has a short, positively charged N-terminal segment and a relatively long C-terminal segment exposed on the cytoplasmic side of the disk membrane. The large 140 amino acid EC-2 domain contains a single N-linked glycosylation site (N229) and seven conserved cysteine residues, six of which are important for protein folding, presumably through the formation of intramolecular disulfide bonds (Connell & Molday 1990, Goldberg et al 1998, Molday 1998, Travis et al 1991). By analogy with the disulfide bonding pattern of CD81 tetraspanin protein, the C165 of the invariant tetraspanin CCG motif most likely forms a disulfide bond with C250 of peripherin 2. A second disulfide bond bridges C166 of the CCG motif to C213 in a second cysteine pair found in a subset of tetraspainin proteins (Seigneuret et al 2001). Finally, a third disulfide bond is predicted to link C214 to C222 as illustrated in Fig. 1. The seventh cysteine (C150) in the EC-2 domain of peripherin-2 is not present in other tetraspanin proteins. This cysteine is not critical for the folding of the EC-2 domain of peripherin 2, but instead forms an intermolecular disulfide bond with a C150 residue in another peripherin 2 or ROM1 protein to form higher order oligomeric complexes as discussed below (Goldberg et al 1998, Loewen & Molday 2000).

The relatively long, 60 amino acid, C-terminal domain exposed on the cytoplasmic side of the disk membrane has been implicated in several functions. A 10 amino acid segment close to the C-terminus has been shown to cause fusion of disk and plasma membranes, a process that is suggested to be important in outer

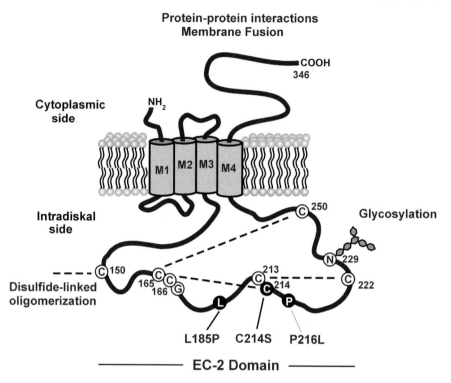

Protein-protein interactions
Membrane Fusion

FIG. 1.   Topological model for Peripherin-2 within the disk membrane. The C150 residue responsible for intermolecular disulfide oligomerization is shown along the three proposed intramolecular disulfide bonds and the location of the conserved N-linked oligosaccharide chain. Also shown are the ADRP mutations (black circles) that have been studied with respect to their effect on subunit assembly, targeting to outer segments, outer segment morphology, and photoreceptor degeneration.

segment renewal (Boesze-Battaglia 2000, Boesze-Battaglia & Goldberg 2002). The C-terminal domain also interacts with other proteins found in rod outer segments (Poetsch et al 2001). Finally, preliminary studies indicate that a region along the C-terminus of peripherin 2 is important in targeting of peripherin 2 to outer segment disk membranes, presumably through interaction with chaperone proteins (Tam 2002).

ROM1 shares an overall amino acid identity with peripherin 2 of over 32% ($\sim$50% in the EC-2 domain) and has a similar topological structure and disulfide bonding pattern (Bascom et al 1992). However, unlike peripherin 2, ROM1 does not contain an N-linked glycosylation site.

Oligomeric structure of peripherin 2 and ROM1

Velocity sedimentation has been used to determine the oligomeric structure of peripherin 2 and ROM1 under non-denaturing conditions (Goldberg & Molday 1996a, Goldberg et al 1995, Loewen & Molday 2000). Peripherin 2 and ROM1 isolated from rod outer segments or transfected COS-1 cells pre-treated with the disulfide-reducing agent dithiotreitol (DTT) co-sediment as core non-covalent tetramers (Fig. 2A). Approximately, two thirds of peripherin 2 exists as homotetramers, while the other third is present as peripherin 2/ROM1 heterotetramers. A small fraction of ROM1, less than 15%, is present as ROM1 homotetramers. This is consistent with the finding that peripherin 2 is over twice as abundant as ROM1 in mammalian rod outer segments.

Under non-reducing conditions, however, peripherin 2 and ROM1 exist as a mixture of core non-covalent tetramers and higher-order disulfide-linked oligomers (Fig. 2B) (Loewen & Molday 2000). Peripherin 2/ROM1 heterotetramers typically assemble as disulfide-linked octomers, whereas peripherin 2 homotetramers form higher-order disulfide-linked oligomers (Fig. 2C). In contrast, ROM1 homotetramers show little tendency to form disulfide-linked complexes (Loewen & Molday 2000). The hetero- and homo-oligomers are held together by C150-mediated disulfide bonds since these oligomeric complexes do not form when the cysteine at position 150 is replaced with a serine residue. The core tetramer, however, is not affected by this cysteine substitution.

The basic unit of the peripherin 2 complex, the core tetramer, exhibits twofold symmetry, and therefore can be considered as a dimer of dimers (Loewen et al 2001). By analogy with CD81, dimerization is most likely mediated by hydrophobic and charged interactions between the A and E helices comprising the stalk region of the EC-2 domain. Tetramerization on the other hand appears to involve residues within the head region since mutations in this part of the EC-2 domain of peripherin 2 and ROM1 abolish tetramer, but not dimer, formation.

Role of peripherin 2 and ROM1 in disk rim structure

Peripherin 2 and ROM1 play distinctive roles in outer segment disk structure and morphogenesis. Peripherin 2 is essential for outer segment disk morphogenesis and structure since outer segments fail to develop in the absence of peripherin 2 (Sanyal & Jansen 1981). Moreover, the level of peripherin 2 is critical for normal outer segment structure, since reduction in peripherin 2 levels by half as in the heterozygous *rds* mouse results in highly disorganized outer segment structures consisting of whorls of disk membranes (Hawkins et al 1984). In contrast, ROM1 appears to play an ancillary role possibly controlling the size and stability

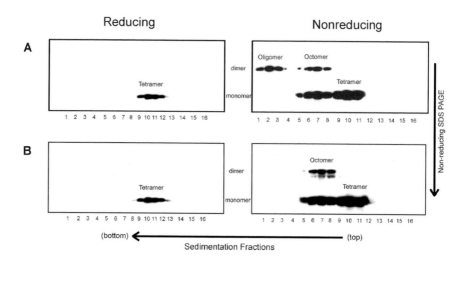

Reducing   Nonreducing

A

B

(bottom) ←————————————————→ (top)

Sedimentation Fractions

C

Core Complexes (a)

Peripherin-2
Homotetramers

Peripherin-2-Rom-1
Heterotetramers

Rom-1
Homotetramers

Intermediate
Disulfide-linked
Octomers (b)

Peripherin-2
Octomers

Peripherin-2:Rom-1
Octomers

Higher ordered
Disulfide-linked
Homo-Oligomers (c)

of the disks. This is based on the finding that *Rom1* knockout mice produce outer segments, the structures of which are only moderately altered (Clarke et al 2000). More specifically, the outer segments of homozygous *Rom1* knockout mice are approximately 25% shorter than those of wild-type mice with relatively mild disorganization of disk stacking. Interestingly, the outer segments also contain some enlarged disks. Heterozygous *Rom1* knockout mice having reduced levels of *Rom1* are normal in appearance (Clarke et al 2000). This suggests that ROM1 may play a role in establishing the preferred size of the disks in mammalian outer segments perhaps — perhaps by modulating the rate of disk closure during outer segment morphogenesis.

The molecular mechanism by which peripherin 2-containing complexes generate and stabilize the outer segments is not well understood. However, several lines of evidence suggest that the oligomeric peripherin 2 complex is important in establishing the high curvature of the disk rim region. The peripherin 2 and ROM1 complexes are specifically localized along the hairpin regions of the rims and incisures of rod and cone disk membranes (Molday et al 1987). Furthermore, *in vitro* expression of wild-type peripherin 2 results microsomes having a flattened disk-like appearance. In contrast expression of C150S peripherin 2 mutant defective in disulfide-linked oligomer formation results in microsomes with typical vesicular structure (Wrigley et al 2000).

How do peripherin 2 oligomers shape the outer segment disk rim? Several models can be envisioned (Fig. 3). In a zipper model, the leaflets of the disks are zippered together through C150 disulfide-linked oligomerization. In this case the core peripherin 2 tetramers are located in adjacent disk bilayers and the disulfide bonds occur across the intradiskal or lumen compartment. In the more logical shell model, peripherin 2-containing oligomers are envisioned to form an oligomeric complex laterally within the membrane and these complexes shape the hairpin curvature of the disk rim. In either case, peripherin 2 and ROM1 appear to reside

---

FIG. 2. (*Opposite*) Analysis of peripherin 2 and ROM1 subunit assembly. (A and B) Velocity sedimentation profiles of reduced and non-reduced complex. Bovine rod outer segment membranes were incubated in the presence (reducing conditions, left panel) or absence (non-reducing conditions, right panel) of 10 mM DTT for 90 minutes. The membranes were then treated with 20 mM N-ethylmaleimide, solubilized in 1% Triton X-100 and subjected to centrifugation on a sucrose gradient. Fractions were collected following the velocity sedimentation run, separated on non-reducing SDS gels and analysed for peripherin 2 (A) and ROM1 (B) by Western blotting. The positions of the tetramers under reducing and non-reducing conditions, and the octomers and higher order oligomers under non-reducing conditions were determined from their sedimentation coefficients. (C) Schematic showing the core non-covalent peripherin 2 and ROM1 homo- and hetero-tetramers and C150-mediated disulfide-linked octomers and higher-order oligomers identified by velocity sedimentation and SDS gel electrophoresis. Figures were modified from Loewen & Molday (2000).

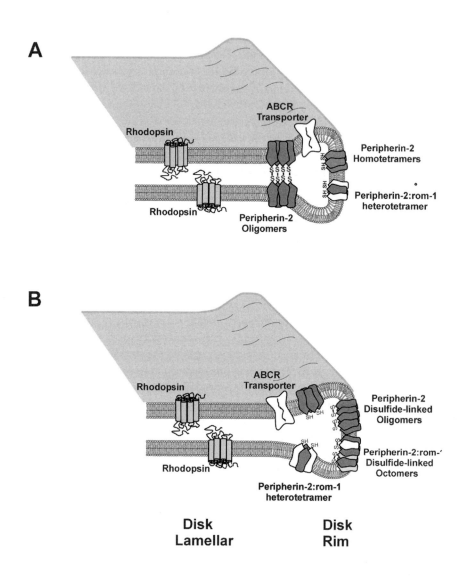

FIG. 3. Models depicting the possible role of peripherin 2/ROM1 complexes in disk rim structure. (A) Zipper model. Peripherin 2 and ROM1 tetramers form disulfide bridges across the disk membranes thereby zippering the leaflets together and generating the hairpin turn of rim region. (B) Shell model. Peripherin 2 homo-oligomers and peripherin 2/ROM1 hetero-oligomers in combination with core tetramers form a supporting shell that establishes the high curvature of the disk rim.

at the disk rim as a mixture of core tetramers and disulfide-linked oligomers. The making and breaking of the C150 intermolecular disulfide bonds possibly catalysed by protein disulfide isomerase may give a degree of plasticity to the membrane curvature during disk rim formation.

In these models, ROM1 is envisioned to play a role in regulating the size of the disks by modulating the content of peripherin 2 oligomeric complexes (Loewen & Molday 2000). In the absence of ROM1, a large fraction of peripherin 2 will be present as higher-ordered disulfide-linked oligomers that may slow the rate of disk closure during the outer segment morphogenesis resulting in enlarged disks. Large disks are found in photoreceptors of lower vertebrates such as *Xenopus laevis* that do not express ROM1 orthologues and in mammalian photoreceptors of *Rom1* knockout mice. In photoreceptor cells expressing ROM1, a significant fraction of peripherin 2 interacts with ROM1 to form octomers that may increase the rate of disk closure resulting in smaller disks as found in mammalian photoreceptors. Further studies are required to more fully assess the role of ROM1 in disk formation.

Peripherin 2/ROM1 complex interacts with the cGMP-gated channel/exchanger and GARP proteins

The interaction of the peripherin 2/ROM1 complex with other rod outer segment proteins has been analysed by co-immunoprecipitation. The cGMP-gated channel, the Na/K–Ca exchanger, GARP proteins, and PDE co-immunoprecipitate with peripherin 2, whereas guanylate cyclase (RetGC1) and ABCR do not (Fig. 4A) (Poetsch et al 2001).

The cGMP-gated channel and the Na/K–Ca exchanger form a complex in the plasma membrane of rod outer segments (Molday & Molday 1998, Schwarzer et al 2000). The channel is a heterotetrameric complex composed of 3 A (or α) subunits and 1 B (or β) subunit (Korschen et al 1995, Molday 1998, Zhong et al 2002). Each subunit contains structural features characteristic of cyclic nucleotide-gated channels including six transmembrane segments, a short pore region and a cyclic nucleotide-binding domain near the C-terminus. Unlike the A subunit, the B subunit of the rod channel contains an extended 571 amino acid N-terminal region known as the glutamic-acid-rich-protein part (or GARP part) (Korschen et al 1995). GARP, in addition to being part of the B subunit of the channel, is also expressed in two soluble isoforms known as GARP1 (or full-length GARP) and GARP2 (or truncated GARP) (Colville & Molday 1996). Bovine GARP1 is identical to the GARP part of the rod B subunit but with an additional 19 amino acids at the C-terminus; GARP2 has the same N-terminal 291 amino acid as GARP1, but a distinct 8 amino acid C-terminal extension.

A

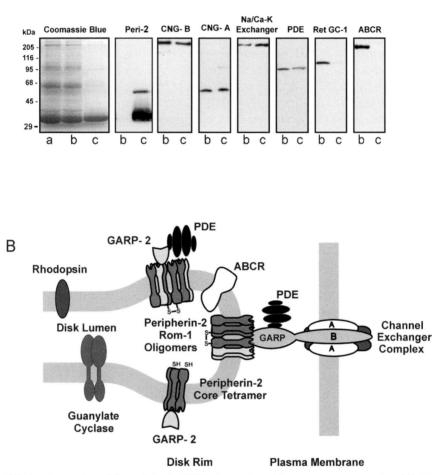

FIG. 4. Interaction of the peripherin 2/ROM1 complex with outer segment proteins. (A) Co-immunoprecipitation of detergent solubilized bovine rod outer segment proteins on an antiperipherin 2–Sepharose matrix. Left panel shows the Coomassie Blue-stained gel of solubilized rod outer segments (lane a), unbound fraction (lane b) and the bound fraction (lane c). The unbound and bound fractions were probed for peripherin 2 (Per-2), the B-subunit (CNG-B) and A-subunit (CNG-A) of the cGMP-gated channel, the Na/K–Ca exchanger, phosphodiesterase (PDE), guanylate cyclase (RetGC1) and ABCR by Western blotting. A significant fraction of the channel subunits, the exchanger and PDE, but not ABCR or RetGC1, co-precipitated with peripherin 2. GARP2 and ROM1 also co-precipitate with peripherin 2 (not shown). (B) A schematic showing the interaction of the peripherin 2:ROM1 oligomers in the disk rim membrane with the channel/exchanger complex in the plasma membrane. This interaction is mediated through the GARP part of the B-subunit of the channel. Peripherin 2/ROM1 complex also interacts with GARP2 and possible GARP1 (not shown). PDE most likely interacts with GARP.

Since the soluble GARP proteins co-immunoprecipitate with peripherin 2, the interaction of the channel/exchanger complex with peripherin 2 appears to be mediated through the GARP part of the channel B-subunit. This protein–protein interaction must span the gap between the disk rim and the plasma membrane since the peripherin 2 complex is localized to the rim region of disks, whereas the channel/exchanger complex is localized to the rod outer segment plasma membrane. Filamentous structures observed under the electron microscope to link the disk rims to the plasma membrane (Roof & Heuser 1982) may reflect this peripherin 2–channel interaction. The peripherin 2–channel interaction may contribute to the formation and stabilization of the unique rod outer segment structure (Fig. 4B). It may also guarantee that the channel/exchanger complex remains in close proximity to the disks to facilitate phototransduction.

The peripherin 2/ROM1 complex also interacts with the soluble GARPs, and in particular the more abundant GARP2 isoform, at the rim region of the disks (Poetsch et al 2001). The role of peripherin 2/ROM1/GARP complex is not known. However, one can speculate that a multimeric GARP2 complex may interact with peripherin 2 on an adjacent disk thereby linking adjacent disks, although this remains to be determined. Fibrous elements have been seen to join adjacent disks by electron microscopy (Roof & Heuser 1982). GARP isoforms also interact with phosphodiesterase (PDE), a key enzyme in the visual cascade (Fig. 4A). This interaction may enable some PDE to remain along the disk rim in close vicinity to the cGMP-gated channel to control cGMP levels. Fig. 4B summarizes protein–protein interactions involving the peripherin 2/ROM1 complex.

Defective subunit assembly of peripherin 2 is responsible for some forms of ADRP

Most missense mutations linked to ADRP are located in the EC-2 domain of peripherin 2. To better understand how selected mutations cause ADRP, we have examined the effect of these disease-linked mutations on the structural properties of peripherin 2 and its interaction with ROM1 using a COS-1 cell expression system in conjunction with immunoprecipitation and sedimentation analysis (Goldberg et al 1998, Goldberg & Molday 1996b). We have also generated transgenic *Xenopus laevis* tadpoles expressing either wild-type or mutant peripherin 2 having a green fluorescence protein (GFP) fused to the C-terminus in order to determine how mutations affect the targeting of peripherin 2 to outer segment disk membranes (Loewen et al 2003).

In the case of C214S peripherin 2 linked to a monogenic form of ADRP (Saga et al 1993), the mutant protein expresses at levels comparable to wild-type peripherin 2 when analysed on disulfide-reducing SDS gels (Goldberg et al

1998). However, instead of assembling as a core tetramer, C214S peripherin 2 forms dimers and aggregates when analysed by velocity sedimentation. The C214S mutation, therefore, prevents the assemblage of dimers to form tetramers, but does not affect dimer formation. This is consistent with the view that the dimerization domain is distinct from the tetramerization domain (Loewen et al 2001). Aggregates observed under non-reducing conditions appear to form as a result of protein misfolding and aberrant intermolecular disulfide bonding. Co-expression studies further indicate that C214S peripherin 2 dimers do not interact with wild-tye peripherin 2 or wild-type ROM1 (Goldberg et al 1998).

We have compared the trafficking of the C214S mutant with wild-type peripherin 2 in rod photoreceptors of transgenic *X. laevis* expressing the transgenes under the control of a rhodopsin promoter (Loewen et al 2003). Whereas wild-type peripherin 2–GFP effectively translocates from the ER of the rod photoreceptor inner segment to the rims and incisures of the outer segment disks, the C214S mutant is retained in the rod inner segment where it localizes to the endoplasmic reticulum, aggresomes and small post-Golgi vesicles near the base region of the cilium (Fig. 5A). Interestingly, the expression of C214S mutant in *X. laevis* does not cause photoreceptor degeneration, at least over a 4 week period. This suggests that the C214S mutant itself does not induce significant degeneration, over the short term. On this basis, we suggest that individuals who inherit the C214S peripherin 2 mutation experience photoreceptor degeneration and ADRP principally as a result of a deficiency in wild-type peripherin 2. Accumulation of C214S in the inner segment, however, may over the long term contribute to the photoreceptor degeneration.

The L185P peripherin 2 mutation linked to a digenic form of ADRP has also been examined in detail. This mutant, like the C214S mutant, assembles as a dimer when individually expressed in COS-1 cells (Goldberg & Molday 1996b, Loewen et al 2001). However, unlike the C214S mutant, L185P peripherin 2 is capable of

FIG. 5.   (*opposite*) Trafficking of wild-type (WT) and C214S peripherin 2–GFP in transgenic *X. laevis*. (A) Fluorescent (left) and electron (right) micrographs of wild-type and C214S peripherin 2–GFP in *X. laevis* retina from 4 week old tadpoles. Wild-type peripherin 2 is effectively transported and incorporated into outer segment disk membranes where is co-localizes with endogenous peripherin 2 to the rims (arrow) and incisures (arrowhead). Inset shows a transverse section showing wild-type peripherin 2–GFP localization to the rim and incisures. C214S is primarily retained in the inner segment where is localizes both to the ER, aggresomes (not shown) and post-Golgi vesicles at the base of the connecting cilium. Occasionally, a small amount of the expressed C214S peripherin 2 is observed at the base of some outer segments, but this protein does not localize to disk membranes. Fluorescent micrographs, bar = 5 μm; electron micrographs, bar = 0.2 μm. (B) Schematic summarizing the effect of peripherin 2 mutations on the subunit assembly, targeting and photoreceptor degeneration. os, outer segment; is, inner segment; cc, connecting cilium (modified from Loewen et al 2003). Subunits with white fill indicate wild-type peripherin 2; subunits in dark fill indicate mutant peripherin 2.

A

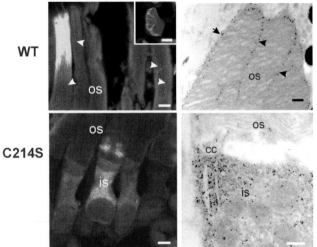

B

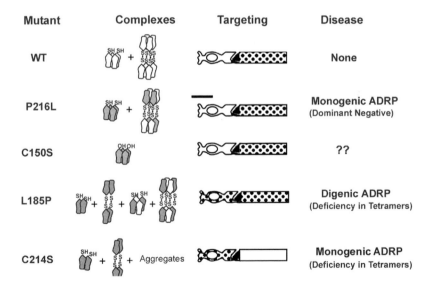

interacting with wild-type peripherin 2 and ROM1 to form core tetramers and higher-order oligomers. This suggests that the L185P mutation causes only localized protein misfolding which prevents two mutant dimers from self-assembling into tetramers. A significant fraction of the mutant, however, can assemble with wild-type peripherin 2 or ROM1 to form native-like tetramers and disulfide-linked oligomers. Hence, the fraction of functional peripherin 2 will be higher in individuals who inherit the L185P mutation compared to those who have the C214S mutation. Hence, individual with the L185P peripherin 2 mutation and normal levels of ROM1 will experience little, if any, photoreceptor degeneration or loss in vision.

Expression of the L185P peripherin 2 mutant in *X. laevis* is consistent with a mixed assembly process. The L185P peripherin 2–GFP protein expressed on an endogenous wild-type peripherin 2 background localizes to both outer segments and inner segments of rod photoreceptors (Loewen et al 2003). The fraction residing in the outer segments most likely represents heterotetramers and disulfide-linked oligomers formed between L185P mutant peripherin 2 and endogenous wild-type protein. The fraction retained in the inner segment consists of the L185P mutant protein that is sufficiently misfolded and incapable of interacting with the wild-type protein. Like the C214S mutant, the L185P mutant does not cause photoreceptor degeneration in *X. laevis* suggesting that normal levels of endogenous peripherin 2 allow for normal photoreceptor disk morphogenesis and cell viability.

These results provide for a molecular basis for digenic ADRP linked to the L185P mutation in peripherin 2 and null or G113E mutation in ROM1. Individuals who inherit a L185P peripherin 2 allele and a null or low expressing G113E ROM1 allele will have levels of functional peripherin 2 containing tetramers and disulfide-linked oligomers below the threshold ($< 70\%$ of normal levels) required for normal outer segment morphogenesis. This will lead to outer segment disorganization and photoreceptor degeneration as found in transgenic mice harbouring these mutations and heterozygous *rds* mice (Hawkins et al 1984, Kedzierski et al 2001). In contrast individuals who inherit only the L185P peripherin 2 mutation, will have sufficient levels of peripherin 2:ROM1 tetramers since a significant portion of L185P can interact with wild-type ROM1 to sustain outer segment disk morphogenesis. As a result, these individuals will show little, if any photoreceptor degeneration or loss in vision. Likewise, individuals who inherit only the ROM1 null allele or the low expressing G113E mutation will have sufficient levels of peripherin 2-containing core tetramers and oligomers for normal outer segment morphogenesis consistent with the heterozygous ROM1 knockout mice (Clarke et al 2000, Loewen et al 2001).

We have also investigated the effect of P216L peripherin 2 ADRP-linked mutation on subunit assembly and peripherin 2 targeting to rod outer segments

(Loewen et al 2003). The bovine mutant appears to assemble as wild-type-like tetramers when expressed by itself and in *X. laevis* transgenic tadpole, the *X. laevis* P216L peripherin 2 mutant targets normally to outer segments. However, unlike the other disease-linked mutants, the P216L mutant causes disorganization of outer segments disks together with photoreceptor degeneration. A similar effect has been described for transgenic mice harbouring the P216L peripherin 2 mutation (Kedzierski et al 1997). The mechanism by which P216L interferes with outer segment morphogenesis and structure is not known, but it has been speculated that it may be due to the introduction of an additional *N*-linked glycosylation in the EC-2 domain. Substitution of proline 216 with a leucine results in a new consensus sequence for *N*-linked glycosylation in mouse, human and *X. laevis* peripherin 2. Indeed, hyperglycosylation of this mutant has been detected by SDS polyacrylamide gel electrophoresis (Wrigley et al 2002, Loewen et al 2003). This additional oligosaccharide chain may affect protein packing and/or protein–protein interactions required for proper disk morphogenesis and stability.

These studies suggest the P216L peripherin 2 mutation, in contrast to the C214S and L185P mutations, exerts a dominant negative effect on outer segment structure resulting in photoreceptor degeneration.

Finally, we examined the effect on the C150S oligomerization deficient mutation on peripherin 2 trafficking to outer segments in *X. laevis* (Loewen et al 2003). In this case, the C150S mutant translocates to outer segment disk membranes indicating that disulfide-linked oligomerization is not required for peripherin 2 transport to outer segments.

Together these studies lead to a model that correlates subunit assembly with peripherin 2 trafficking to outer segment disks and ADRP as shown in Fig. 5B. In general tetramerization, but not higher-order oligomerization, is required for translocation of peripherin 2 to outer segments. ADRP linked to C214S and L185P mutations are caused by a deficiency in peripherin 2-containing tetramers, whereas ADRP linked to the P216L mutation appears to result from a dominant negative effect of the mutant on disk morphogenesis and structure.

Conclusions

Peripherin 2 together with ROM1 assemble as core tetramers which can further link together to form higher-ordered disulfide-linked oligomers. These complexes are important in disk morphogenesis and structure as well as interaction with GARP-containing rod outer segment proteins including the channel/exchanger complex in the plasma membrane. Tetramer formation is critical for the trafficking of peripherin 2 from the inner segment through the cilium and to the rims and incisures of outer segment disks. ADRP mutations that affect tetramerization such as C214S and L185P cause photoreceptor degeneration principally through a deficiency of wild-type peripherin 2-containing tetramers and oligomers, whereas the P216L ADRP mutation causes photoreceptor degeneration through a dominant negative mechanism possibly involving the introduction of a second N-linked oligosaccharide chain into the EC-2 domain of peripherin 2. Studies indicate that the peripherin 2:ROM1 complex is a multi-functional protein critical for proper outer segment morphogenesis. More detailed analysis of the function and transport of peripherin 2 and ROM1 will be important in further defining the role of this complex in outer segment morphogenesis and retinal degenerative diseases. It will be of particular interest to:

- define the molecular mechanism by which the peripherin 2:ROM1 complex mediates disk rim formation
- further identify distinct functions of the peripherin 2:ROM1 complex in rod and cone photoreceptors, and
- elucidate the mechanism by which specific peripherin 2 mutations lead to diverse retinal disease phenotypes.

Acknowledgements

This research was supported by grants from the National Institutes of Health (EY 02422), the Canadian Institutes for Health Research (MT 5822), and the Foundation Fighting Blindness — Canada.

References

Arikawa K, Molday LL, Molday RS, Williams DS 1992 Localization of peripherin/rds in the disk membranes of cone and rod photoreceptors: relationship to disk membrane morphogenesis and retinal degeneration. J Cell Biol 116:659–667

Bascom RA, Manara S, Collins L, Molday RS, Kalnins VI, McInnes RR 1992 Cloning of the cDNA for a novel photoreceptor membrane protein (rom-1) identifies a disk rim protein family implicated in human retinopathies. Neuron 8:1171–1184

Boesze-Battaglia K 2000 Fusion between retinal rod outer segment membranes and model membranes: functional assays and role for peripherin/rds. Methods Enzymol 316:65–86

Boesze-Battaglia K, Goldberg AF 2002 Photoreceptor renewal: a role for peripherin/rds. Int Rev Cytol 217:183–225

Cheng T, Peachey NS, Li S, Goto Y, Cao Y, Naash MI 1997 The effect of peripherin/rds haploinsufficiency on rod and cone photoreceptors. J Neurosci 17:8118–8128

Cideciyan AV 2000 In vivo assessment of photoreceptor function in human diseases caused by photoreceptor-specific gene mutations. Methods Enzymol 316:611–626

Clarke G, Goldberg AF, Vidgen D et al 2000 Rom-1 is required for rod photoreceptor viability and the regulation of disk morphogenesis. Nat Genet 25:67–73

Colville CA, Molday RS 1996 Primary structure and expression of the human beta-subunit and related proteins of the rod photoreceptor cGMP-gated channel. J Biol Chem 271:32968–32974

Connell G, Bascom R, Molday LL, Reid D, McInnes RR, Molday RS 1991 Photoreceptor peripherin is the normal product of the gene responsible for retinal degeneration in the rds mouse. Proc Natl Acad Sci USA 88:723–726

Connell GJ, Molday RS 1990 Molecular cloning, primary structure, and orientation of the vertebrate photoreceptor cell protein peripherin in the rod outer segment disk membrane. Biochemistry 29:4691–4698

Dryja TP, Hahn LB, Kajiwara K, Berson EL 1997 Dominant and digenic mutations in the peripherin/RDS and ROM1 genes in retinitis pigmentosa. Invest Ophthalmol Vis Sci 38:1972–1982

Farrar GJ, Kenna P, Jordan SA et al 1991 A three-base-pair deletion in the peripherin-RDS gene in one form of retinitis pigmentosa. Nature 354:478–480

Goldberg AF, Molday, RS 1996a Subunit composition of the peripherin/rds-rom-1 disk rim complex from rod photoreceptors: hydrodynamic evidence for a tetrameric quaternary structure. Biochemistry 35:6144–6149

Goldberg AF, Molday RS 1996b Defective subunit assembly underlies a digenic form of retinitis pigmentosa linked to mutations in peripherin/rds and rom-1. Proc Natl Acad Sci USA 93:13726–13730

Goldberg AF, Moritz OL, Molday RS 1995 Heterologous expression of photoreceptor peripherin/rds and Rom-1 in COS-1 cells: assembly, interactions, and localization of multisubunit complexes. Biochemistry 34:14213–14219

Goldberg AF, Loewen CJ, Molday RS 1998 Cysteine residues of photoreceptor peripherin/rds: role in subunit assembly and autosomal dominant retinitis pigmentosa. Biochemistry 37:680–685

Hawkins RK, Jansen HG, Sanyal S 1984 Development and degeneration of retina in rds mutant mice: photoreceptor abnormalities in the heterozygotes. Exp Eye Res 41:701–720

Hemler ME 2001 Specific tetraspanin functions. J Cell Biol 155:1103–1107

Jacobson SG, Cideciyan AV, Maguire AM, Bennett J, Sheffield VC, Stone EM 1996 Preferential rod and cone photoreceptor abnormalities in heterozygotes with point mutations in the RDS gene. Exp Eye Res 63:603–608

Kajiwara K, Hahn LB, Mukai S, Travis GH, Berson EI, Dryja TP 1991 Mutations in the human retinal degeneration slow gene in autosomal dominant retinitis pigmentosa. Nature 354:480–483

Kajiwara K, Berson EL, Dryja TP 1994 Digenic retinitis pigmentosa due to mutations at the unlinked peripherin/RDS and ROM1 loci. Science 264:1604–1608

Kedzierski W, Lloyd M, Birch DG, Bok D, Travis GH 1997 Generation and analysis of transgenic mice expressing P216L-substituted rds/peripherin in rod photoreceptors. Invest Ophthalmol Vis Sci 38:498–509

Kedzierski W, Nusinowitz S, Birch D et al 2001 Deficiency of rds/peripherin causes photoreceptor death in mouse models of digenic and dominant retinitis pigmentosa. Proc Natl Acad Sci USA 98:7718–7723

Kitadokoro K, Bordo D, Galli G et al 2001 CD81 extracellular domain 3D structure: insight into the tetraspanin superfamily structural motifs. EMBO J 20:12–28

Korschen HG, Illing M, Seifert R et al 1995 A 240 kDa protein represents the complete beta subunit of the cyclic nucleotide-gated channel from rod photoreceptor. Neuron 15:627–636

Loewen CJ, Molday RS 2000 Disulfide-mediated oligomerization of Peripherin/Rds and Rom-1 in photoreceptor disk membranes. Implications for photoreceptor outer segment morphogenesis and degeneration. J Biol Chem 275:5370–5378

Loewen CJ, Moritz OL, Molday RS 2001 Molecular characterization of peripherin-2 and rom-1 mutants responsible for digenic retinitis pigmentosa. J Biol Chem 276:22388–22396

Loewen CJ, Moritz OL, Tam BM, Papermaster DS, Molday RS 2003 The role of subunit assembly in peripherin-2 targeting to rod photoreceptor disk membranes and retinitis pigmentosa. Mol Biol Cell 14:3400–3413

Molday RS 1998 Photoreceptor membrane proteins, phototransduction, and retinal degenerative diseases. The Friedenwald Lecture. Invest Ophthalmol Vis Sci 39: 2491–2513

Molday RS, Hicks D, Molday L 1987 Peripherin. A rim-specific membrane protein of rod outer segment discs. Invest Ophthalmol Vis Sci 28:50–61

Molday RS, Molday LL 1998 Molecular properties of the cGMP-gated channel of rod photoreceptors. Vision Res 38:1315–1323

Moritz OL, Molday RS 1996 Molecular cloning, membrane topology, and localization of bovine rom-1 in rod and cone photoreceptor cells. Invest Ophthalmol Vis Sci 37: 352–362

Poetsch A, Molday LL, Molday RS 2001 The cGMP-gated channel and related glutamic acid-rich proteins interact with peripherin-2 at the rim region of rod photoreceptor disc membranes. J Biol Chem 276:48009–48016

Roof DJ, Heuser JE 1982 Surfaces of rod photoreceptor disk membranes: integral membrane components. J Cell Biol 95:487–500

Saga M, Mashima Y, Akeo K, Oguchi Y, Kudoh J, Shimizu N 1993 A novel Cys-214-Ser mutation in the peripherin/RDS gene in a Japanese family with autosomal dominant retinitis pigmentosa. Hum Genet 92:519–521

Sanyal S, Jansen, HG 1981 Absence of receptor outer segments in the retina of rds mutant mice. Neurosci Lett 21:23–26

Schwarzer A, Schauf H, Bauer PJ 2000 Binding of the cGMP-gated channel to the Na/Ca-K exchanger in rod photoreceptors. J Biol Chem 275:13448–13454

Seigneuret M, Delaguillaumie A, Lagaudriere-Gesbert C, Conjeaud H 2001 Structure of the tetraspanin main extracellular domain. A partially conserved fold with a structurally variable domain insertion. J Biol Chem 276:40055–40064

Tam BM, Moritz OL, Hurd LB, Papermaster DS 2002 Characterization of the functional properties of the C-terminus of Xenopus Peripherin [ARVO Abstract]. Invest Ophthalmol Vis Sci 43:B729 (abstr 3745)

Travis GH, Brennan MB, Danielson E, Kozak CA, Sutcliffe JG 1989 Identification of a photoreceptor-specific mRNA encoded by the gene responsible for retinal degeneration slow (rds). Nature 338:70–73

Travis GH, Sutcliffe JG, Bok D 1991 The retinal degeneration slow (rds) gene product is a photoreceptor disc membrane-associated glycoprotein. Neuron 6:61–70

Weleber RG, Carr RE, Murphey WH, Sheffield VC, Stone EM 1993 Phenotypic variation including retinitis pigmentosa, pattern dystrophy, and fundus flavimaculatus in a single family with a deletion of codon 153 or 154 of the peripherin/RDS gene. Arch Ophthalmol 111:1531–1542

Wrigley JD, Ahmed T, Nevett CL, Findlay JB 2000 Peripherin/rds influences membrane vesicle morphology. Implications for retinopathies. J Biol Chem 275:13191–13194

Wrigley JD, Nevett CL, Findlay JB 2002 Topological analysis of peripherin/rds and abnormal glycosylation of the pathogenic Pro216→Leu mutation. Biochem J 368: 649–655

Zhong H, Molday LL, Molday RS, Yau KW 2002 The heteromeric cyclic nucleotide-gated channel adopts a 3A:1B stoichiometry. Nature 420:193–198

DISCUSSION

Bok: Could you clarify one thing? Was the central versus peripheral P216L effect transient, and did it spread to the periphery over time as the frog got older? Or was it persistent?

Molday: The animals that we analysed were quite young — on average 4–8 weeks old. For P216L we only examined 4 week-old animals and we only saw degeneration in the central retina. We anticipate that the degeneration will spread to the peripheral region of the retina as the animals get older. Unfortunately, in this kind of experiment you have to sacrifice the animal. We did not generate enough transgenic *X. laevis* tadpoles to examine the temporal and spatial profile for degeneration.

Bok: Do you have a line going or will you have to make new animals?

Molday: We didn't make a line. We hope to generate additional mutant tadpoles and make a line in the future to examine the time dependence of degeneration.

LaVail: I want to make sure I understood you. You said you went up to about eight weeks in all of them. Would you think that with these kinds of changes piling up in the inner segment, you would ultimately get some changes in photoreceptors?

Molday: We did look at some of the animals up to 8 weeks. The wild-type transgenic animals looked quite normal, although we did not carry out a detailed study. I would suspect that the accumulation of the C214S dimeric species in the inner segment over time would cause problems, eventually leading to degeneration. In people with the C214S mutation, however, I would predict that photoreceptor degeneration is principally caused by low levels of wild-type peripherin.

McInnes: Have you seen any interaction between either of these proteins and ABCR?

Molday: No. We carried out extensive immunprecipitation and cross-linking studies and find no co-precipitation or cross-linking of ABCR with peripherin 2 or ROM1. This is a negative experiment. It is possible that detergent solubilization could disrupt existing weak ABCR–peripherin 2 interactions and cross-linking could be ineffective. At the present time, however, we have no evidence to suggest that ABCR strongly interacts with peripherin 2.

McInnes: We observed larger disks in the *Rom1* mice, and Dean Bok has pointed out that the *Rds* heterozygote mice have large disks as well. How do you think that phenotype fits into the biochemical model you are proposing?

Molday: It fits quite well. Transgenic *Xenopus* tadpoles expressing wild-type transgene against a wild-type background express peripherin in excess over normal levels. At high transgenic peripherin expression we have seen a narrowing of the outer segment and smaller disks. This appears to indicate that excess peripherin 2 increases the rate of disk closure resulting in smaller disks.

When peripherin or ROM1 is in lower than normal amounts such as in the *Rom1* knockout mouse and *Rds* heterozygote mouse, disk rim closure is slow resulting in larger disks. Interestingly, amphibians such as *X. laevis* with large disks do not have ROM1. This implies that ROM1 may be a negative regulator of disk size via its interaction with peripherin 2.

Hauswirth: In the peripherin mutations in which you see accumulation in the inner segment, have you stained with antibodies against the normal *Xenopus* proteins such as rhodopsin, to see whether there is some effect on general transport to the outer segments?

Molday: We have examined endogenous wild-type peripherin 2 in the C214S transgenics that accumulate the mutant in the inner segment. In this instance, the trafficking of endogenous wild-type peripherin 2 to the outer segment is not affected by the presence of C214S accumulation in the inner segments. We have not looked at other outer segment proteins. Since outer segments appear normal in these animals, we assume that the trafficking of other outer segment proteins is normal. However, this is a reasonable experiment to do. We should examine protein trafficking to outer segments in these mutant animals by immunofluorescence microscopy.

Daiger: There are three different polymorphic amino acid substitutions in human peripherin 304, 310 and 338, which lead to four different protein haplotypes. Have you looked for functional differences in those four haplotypes? And in your constructs for dealing with human mutations, do you pick only one of those four haplotype backgrounds?

Molday: Our studies have concentrated on disease-linked mutations. We have not examined the polymorphic amino acid substitutions. We have no evidence that these changes will affect the biochemical properties or subunit assembly of peripherin 2.

Dryja: There is no evidence that those polymorphisms are related to any human photoreceptor disease.

Daiger: No. I'm just asking whether the four different proteins produced by that locus are functionally relevant.

Dryja: One piece of evidence in favour of them not being relevant is that the fluorescent tag is placed on the end of the protein that has the polymorphic residues. Even with the tag on that end of the protein, the protein seems to work fine. It looks like that end of the protein is not functionally important.

Bok: *Xenopus* actually has three different peripherin genes, leading to three different peripherin proteins in the outer segment.

Molday: We have concentrated on the major form that is most similar to human peripherin 2. I believe the two other isoforms are present in lower concentrations.

Bok: All three of the isoforms of peripherin/rds in *Xenopus* are readily detectable by immunocytochemistry, so they must be fairly abundant.

Dryja: I'd like to follow up this question about a possible interaction between ABCR and peripherin. One blot showed no evidence of an interaction, but if there were some way that the peripherin mutations interfered with ABCR activity, it might explain why we get these macular degeneration phenotypes with subretinal deposits that Alan Bird was mentioning that are reminiscent of the macular degeneration that is associated with ABCR mutations.

Molday: This would require a co-expression system in which one could co-express wild-type and mutant peripherin with wild-type ABCR and look for differences in the function of ABCR at a biochemical level. A system could be developed to examine the ATPase activity of ABCR in the presence of wild-type and mutant peripherin. This has not yet been done.

Travis: Benjamin Kaupp showed an interaction between free GARP and ABCR. Although there may not be a direct interaction between Rds and ABCR, they may both be talking through GARP.

Molday: The study by Kaupp and co-workers showing an interaction between ABCR and free GARP has been retracted (see the recent review article Kaupp & Seifert 2002). It would appear that the interaction they initially described was due to non-specific interactions.

Nathans: I have two questions that might be related. What do you think controls the generation of incisures? You can imagine with this linkage to the plasma membrane how peripherin/rds complexes might assemble, but what makes them move into the interior of a disk to form an incisure, and what controls the number of incisures? And what is free GARP doing?

Molday: It seems that the number of incisures is often greater in larger animals. I do not know what the molecular or cellular mechanisms that control the number and formation of incisures. This, along with the molecular basis of disk formation, is poorly understood at the present time.

I can speculate on the possible role of free GARP. We know that GARP on the channel and free GARP bind to the peripherin/ROM1 complex at the rim region of disk membranes. The interaction of the GARP part of the B-subunit of the channel may be largely responsible for defining the spatial relations of the disks to the plasma membrane via this protein–protein interaction and may represent filamentous connections observed between the disk rim and plasma membrane by electron microscopy. One can suggest that free GARP may control the spatial arrangement of the stack of disks. If GARP forms a multi-subunit complex, then this complex can be envisioned to form a bridge between adjacent disks via interactions with the peripherin/ROM1 complex at the rim regions of disks. Therefore, GARP with a high content of proline residues would bind peripherin/ROM1 and link together adjacent disk rims and disks and plasma membrane via protein–protein interactions involving peripherin–GARP complexes.

Nathans: Has free GARP been immunolocalized?

Molday: Yes. GARP has been localized along the rim regions of disk membranes and between the disks and plasma membrane using both post-embedding and pre-embedding immunogold labelling techniques.

Bok: You showed a transmission electron micrograph that displayed filaments going from neck to neck in the outer segment disk loops. Do you think this is free GARP?

Molday: That's our speculation. Other proteins may also be involved although the number of unidentified candidate proteins in outer segments that are of sufficient quantity to represent the filaments observed between two disks and between a disk and plasma membrane is diminishing.

Aguirre: If I extend the free GARP question, in RPGR exon ORF15 there is a very glutamic acid-rich region which is almost GARP-like. Can you foresee a GARP-like role for RPGR in that region of the outer segment?

Molday: The GARP nomenclature is a bit of a misnomer. When GARP1 was identified it was characterized as having a string of glutamic acids near the C-terminal part of the protein. This is also present in the GARP part of the channel. GARP2, a truncated form of GARP1, however, is the predominant form in the outer segments and does not have this glutamic acid-rich region. In fact, it is a proline-rich protein as opposed to a glutamic acid-rich protein. My speculation would be that this glutamic acid-rich region, which is not well conserved, probably functions to extend the functional GARP domain further out thereby facilitating the interaction between the disc and the plasma membrane.

Reference

Kaupp UB, Seifert R 2002 Cyclic nucleotide-gated ion channels. Physiol Rev 82:769–824

The search for rod-dependent cone viability factors, secreted factors promoting cone viability

Thierry Léveillard, Saddek Mohand-Saïd, Anne Claire Fintz, George Lambrou*, José-Alain Sahel[1]

*Laboratoire de Physiopathologie Cellulaire et Moléculaire de la Rétine, Université Pierre et Marie Curie and Inserm u592, Hôpital Saint-Antoine, Bâtiment Kourilsky, 6ème étage, 184 rue du Faubourg Saint-Antoine, 75571 Paris Cedex 12, France and *Novartis-Ophthalmics, Klybeckstrasse 141, CH-4002 Basel, Switzerland*

Abstract. During the last decade, numerous research reports have considerably improved our knowledge of the pathophysiology of retinal degenerations. Three non-mutually exclusive general areas dealing with therapeutic approaches have been proposed: gene therapy, pharmacology and retinal transplantations. The observation that cone photoreceptors, even those seemingly unaffected by any described anomaly, die secondarily to rod disappearance related to mutations expressed specifically in the latter, led us to study the interactions between these two photoreceptor populations to search for possible causal links between rod degeneration and cone death. These *in vivo* and *in vitro* studies suggest that paracrine interactions between both cell types exist and that rods are necessary for continued cone survival. We have developed a protocol that is used to evaluate the potential of all sequences in a retinal library to generate a protective effect on cones from cone-enriched cultures from chicken embryo. The protocol of expression cloning is a systematic approach aimed at screening all genes normally expressed by retina. Since the role of cones in visual perception is essential, pending the identification of the factors mediating these interactions underway, rod replacement by transplantation and/or neuroprotection by trophic factors or alternative pharmacological means appear as promising approaches for limiting secondary cone loss in currently untreatable blinding conditions.

2004 Retinal dystrophies: functional genomics to gene therapy. Wiley, Chichester (Novartis Foundation Symposium 255) p 117–130

The foremost cause of irreversible blindness in major retinal diseases is photoreceptor degeneration. In animal models as well as in human retinal hereditary dystrophies, the mutations described since 1990 affect mainly coding

---

[1]This paper was presented at the symposium by José-Alain Sahel. Correspondence should be addressed to Thierry Léveillard or José-Alain Sahel.

sequences for structural proteins (peripherin, ROM1) or components of the phototransduction cascade (rhodopsin, cGMP-dependent phosphodiesterase) found in the rod outer segments (Rosenfeld et al 1992, McLaughlin et al 1993, Kajiwara et al 1994). The mechanisms leading to programmed cell death of these cells are still hypothetical (Chang et al 1993). In addition to this direct rapid rod loss, delayed cone loss is seen in clinical situations and was described in 1978 in the 'retinal degeneration' (rd) mouse model by Carter-Dawson (1978). This secondary cell loss does not have any obvious explanation since cones are generally not directly affected by the genetic anomaly found in these diseases. Their loss is responsible for the major visual handicap because cones are essential for diurnal, colour and central vision (Dowling 1987).

In several models leading to selective rod loss, such as transgenic mice (McCall et al 1996) or mice carrying a spontaneous mutation (Bowes et al 1990), secondary cone loss is observed whereas the causal abnormality is not directly incriminated in their degeneration. In certain studies the link between rod loss and cone dropout is still hypothetical. The cellular interactions involved in cone survival have never been the subject of a systematic experimental approach, and can be amply justified through the major perspectives in fundamental neurobiology and therapeutic outcomes. Taking into account the multiple cone functions, preservation of this population would open an original avenue of therapeutic investigation which would enable a considerable limitation of functional consequences for the patients.

Potential applications include the incurable pathologic group collectively known as age-related macular degeneration (AMD), current treatments for which are still palliative and relatively inefficient. In AMD, the most widely accepted hypothesis proposes that degeneration results from a defect in photoreceptor (PR) renewal by the retinal pigmented epithelium (RPE). The disks which compose the PR outer segments are renewed daily at a high sustained rhythm. This renewal is carried out both by new protein synthesis and by elimination of outer segment debris by the RPE. With aging this renewal becomes increasingly inefficient, leading to progressive accumulation of fragments called drüsen. Simultaneous to this accumulation, choroidal atrophy and/or neovascular invasion leading to irreversible destruction of PR and RPE occurs. As only a few cases (about 15%) are able to be treated palliatively by laser photocoagulation, AMD is the principal cause of visual impairment in western countries.

The pathogenic mechanisms involved in this condition are still hypothetical. If the prevailing opinion situates the initial mechanism at the level of the phagocytosis defect in RPE, several arguments indicate that rod malfunction could represent the initial site of lesion. Among the most frequent hypotheses, phototoxicity is considered as a major mechanism. Whereas exposure to strong light intensities is often responsible for RPE alterations, Noell (1980) showed that at ordinary intensities, lesions are found within the PRs. In these cases, effects cumulate with very low thresholds compatible with infraliminal attacks occurring throughout life. Curcio et al (1993, 1996) demonstrated early specific rod loss in retinas from donor eyes prior to visible macular alteration, even though these eyes showed drüsen accumulation or pigmentary changes. Furthermore, these authors established that at more advanced AMD stages, rod loss occurs earlier and in greater abundance than that of cones. Rod loss occurs in the parafoveal region where a high cell density is found, and correlates with data obtained from functional tests such as obscurity adaptation curves and central microperimeter measurements (Sunness et al 1989, Owsley et al 2000). Early rod cell death might therefore be envisioned as an intermediate mechanism leading to loss of central vision and as a target for early intervention.

Rod–cone interaction

Hereditary retinal dystrophies currently offer the best approach to the identification of pathways, targets and molecules involved in PR (and particularly cone) cell degeneration and protection using knowledge gained through the identification of many disease-causing mutations as well as the analysis and development of relevant animal models. In many of these conditions rod loss through apoptosis occurring in a cell autonomous manner is followed by cone degeneration. Cones are expressing the mutant gene in only a minority of conditions, and degenerate therefore in a non-cell autonomous mechanism. Since cones are the class of PRs most important for vision, we designed strategies aimed at protecting these cells, in order to propose a therapeutic approach that would alleviate a common mechanism leading to blindness in several different conditions affecting initially rods, independent from the nature of the causative mutation.

Using a mouse model of human recessive RP, the *rd1* mouse (loss of function of PDE6B), our group has demonstrated that transplantation of sheets of pure PRs (97% rods) isolated from a wild-type animal into the subretinal space of the rodless *rd1* mouse delay significantly the secondary loss of cones (Mohand-Saïd et al 1997, 2000). The effect is specifically mediated by the rods since inner retina from a wild-type animal as well as whole retina from a rodless animal (8 weeks *rd1* retina) nor the gelatin substrate of transplanted cells are mediating the 40% survival observed with PR sheets (Fig. 1). The effect, observed at distance from the grafted tissue led us to propose the hypothesis that the survival activity is mediated by molecules secreted by rods (or only in the presence of them). We validated this hypothesis using an *in vitro* assay recapitulating some of the paradigm of transplantation (Mohand-Saïd et al 1998) in these co-cultures, in which the samples are separated by a filter that allows only communication through soluble molecules. The

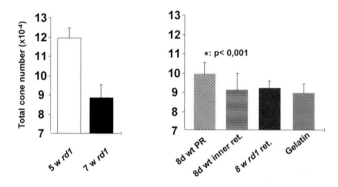

FIG. 1.   Survival of cones after transplantation of sheets of PRs.

neuroprotective effect on cones was measured using a specially designed method (stereology counting) that very significantly decreases the standard deviation, allowing accurate estimates of the total numbers of cells. Wild-type 8 d as well as 5 weeks normal neural retinas are mediating 40–50% increase in cone viability in *rd1* retinal explants, while the rodless 5 weeks *rd1* retina is devoid of this activity (Fig. 2). The viability activity is also carried by *rd1* retina, before the degeneration of rods (at 8 days) indicating that the mutation affecting rod function in rods does not directly alter the communication between the two classes of PR. We also established that conditioned media (CM) from wild-type retina contains activity that promote cone viability that is heat sensitive (Fintz et al 2003). This activity has an apparent molecular weight larger than 25 kDa (Fig. 3). The heat sensitivity and the apparent molecular weight indicate that the activity is carried

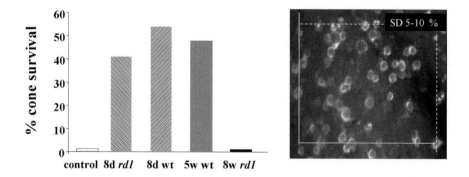

FIG. 2.   Cones survival is mediated by rod-dependent secreted molecule.

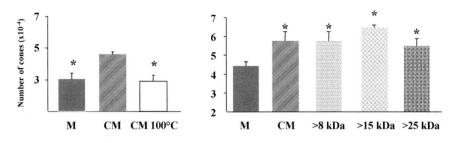

FIG. 3.   The rod-dependent soluble cone survival activity is heat sensitive and has an apparent molecular weight larger than 25 kDa.

out by protein(s) termed the rod-dependent cone viability factor(s). We postulate that the degeneration of rods of the *rd1* mouse retina is leading to the loss of expression of secreted protein factor(s) essential for cone viability. This mechanism of cone degeneration is also likely in human retinas affected with RP, at a stage when most rods have degenerated, before central vision is under threat (data from post-mortem and functional studies, e.g. Cideciyan et al 1998) .The identification of the genes encoding these factor(s) is a prerequisite to a replacement therapy aimed at preventing the secondary loss of cones and of vision.

Identification of rod-dependent cone viability factors

The main challenges in the identification of RdCVFs are that it requires the use of a functional assay that should be reproducible and easy. The methods for estimating the number of cones after labelling with a specific lectin (PNA) is accurate but due to the fact that cones represent only 3% of the PR in rodents and due to the fact that the actual count is estimated by stereology on flat-mounted retinal explants, it is time consuming. The cone-enriched culture system we developed is based on the work published by Adler. To the opposite of the mammals, birds have retinas dominated by cones. Isolation of retinal precursors from embryonic chicks retinas leads to their differentiation into PR by a default pathway when isolated from any developmental cues (Adler & Hatlee 1989). In these cultures, in the absence of serum, the cones represent 60–80% of the cells. These primary postmitotic cells are degenerating over a period of 9 days (M in Fig. 5). We have developed polyclonal antibodies against visinin, a PR marker and confirmed that this is a cone-enriched culture system (Fig. 4). The morphology of these cells is characterized by a liquid droplet and their bipolarity so that the number of cone-like cells in a culture dish can be measured. Prior to the use of the functional assay to isolate RdCVFs we checked the degeneration kinetics of the cones in the presence

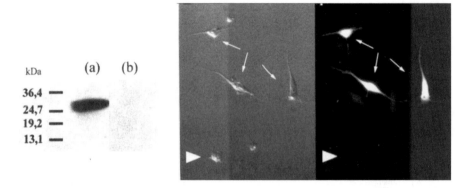

FIG. 4.   Chicken retinal cultures cells labelled with polyclonal antibodies raised against visinin. Left: western blotting analysis of chick retinal extracts (a) immune and (b) preimmune serum. Middle: Nomarski. Right: immunofluorescence of chicken retinal cultures labelled with the visinin antibodies. Arrows, cones; arrowhead, other retinal cell.

of CM isolated from wild-type mouse retina (Fintz et al 2003). The viability was recorded using a metabolic activity (Fig. 5, top panel) as well as using cell counting (bottom panel). In both cases we observed an increase in cell survival after a period extending from 5 to 9 days of culture. This validation of the effect of RdCVF activity on cones isolated from chicken embryo made this functional assay easy, reproducible and high throughput. We determined that the viability activity on chicken cone is heat labile and measured its apparent molecular weight using serial filtration of the mouse conditioned media. Part of the activity has an apparent molecular weight larger than 25 kDa as per the activity measured on mouse cones. The decrease in activity when removing molecules inferior to 15 kDa indicates that several species are mediating the effect on chicken cones, only part of it corresponding to the mouse RdCVF activity (Fintz et al 2003).

    The protocol of expression cloning we used to identify the RdCVF genes is a systematic approach aimed at screening for all the chicken cone promoting activities expressed by wild type neural retina using the chicken cone system. Thereafter, a major step is to validate their therapeutic potential using the *rd1* mouse explants.

    The first step was the construction of an expression library from wild-type adult mouse adult retina (C57BL/6) in the plasmid vector pcDNA3. This oriented library was constructed using published protocol and was used in the experiment without any amplification in order to avoid any loss of non-abundant clones.

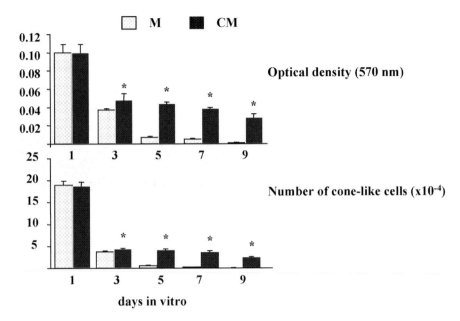

FIG. 5.   RdCVF activity on cone-enriched cultures.

Plasmids from the library were pooled by 100 clones and purified using QIA well 96 kit. We have documented using our functional assay and the protocol of transient transfection that the activity of glial-derived neurotrophic factor (GDNF) on chicken cone cells is detected at a 1/100 plasmid dilution (data not shown). The plasmid pools were transfected into a cell line (COS-1) that is easily and reproducibly transfected by the calcium phosphate method at high efficiency. The COS-1 cells have been created by transformation with the SV40 large T antigen that allows replication of the plasmid through the SV40 origin of replication (SV40 Ori). In addition to the strong CMV promoter, this protocol allows high expression of the cDNA. After transfection, the conditioned media from the COS-1 transfected cells (cultured 2 days without serum) is harvested and incubated with primary chicken cone cells for 7 days (according to Fig. 5). The viable cells from the cone-enriched cultures are counted and compared to those of cultures with conditioned media from COS-1 cells transfected with the empty vector pcDNA3. After statistical analysis, the pools generating a positive effect are subdivided in smaller pools of 10 that are tested and the isolation of single cDNA followed by limited dilution (Fig. 6). Using this protocol, the primary screen proceeds at a rate of 8000 clones (80 pools of 100) a week. In

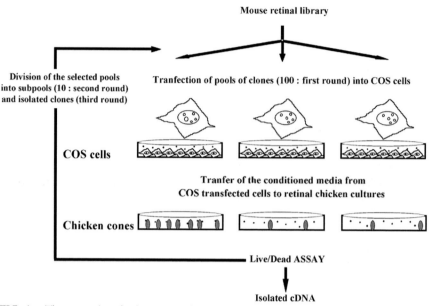

FIG. 6.   The expression cloning protocol.

order to monitor the viability of these cultures, we used a viability assay based on fluorogenic probes and developed specific algorithms on the software Metamorph, designed in order to acquire images on an inverted microscope and automatically count the viable cells. To deal with the variability inherent to the culture of primary cells, each pool was tested in a total of 4 wells of two 96-well plates at slightly different seeding densities. The images were recorded; the cell counting was done for each plate with morphometric parameters (area of a single cell, etc.) calculated for each experiment. The data were expressed as fold difference with the control. In the first part of our protocol, 2100 pools (a total of 210 000 clones) were screened, and from these 24 pools were selected for further analysis. The pool 939 was prioritized based on its activity and we further proceeded to the isolation of a single sequence. The isolated clone contains an open reading frame.

After having screened more than 200 000 clones (2000 pools) we have realized that the bottleneck in our protocol is, after the screening on chick embryo retinal cells, the validation using the mouse model of RP, the *rd1* mouse. We plan to set up an assay based on cells labelled with the visinin antibodies (direct counting of cones). Initially, the cell counting on mouse retina was done using semi-automated methods. We have developed an automated acquisition and cell counting of cones on mouse retinal explants as well as generated transgenic mice with fluorescent cones.

Perspectives and conclusions

The discovery of the neuroprotective effect of the growth factor fibroblast growth factor (FGF)2 in a rat model of inherited retinal dystrophy (Faktorovich et al 1990) was at the origin of a therapeutic strategy aiming to administer trophic factors. It was soon recognized that while FGF2 has a beneficial effect on PRs it couldn't be used in a treatment of retinal degeneration due to its property of triggering neovascularization (Perry et al 1995). The search for a more appropriate trophic factor was pursued and, by evaluating the known factors, two different groups have identified the ciliary neurotrophic factor (CNTF, Cayouette & Gravel 1997, LaVail et al 1998) as well as our demonstration of the activity of GDNF (Frasson et al 1999). Electroretinogram (ERG) recording of the treated animals reveals that GDNF has a benefit for the function of the PRs while a decrease in ERG amplitude was observed using CNTF (Liang et al 2001). The question of the preservation of the PR function is central to any therapeutic strategy. The basis for these decreased ERG amplitudes may be related to changes in gene expression (Bok et al 2002) and to the ability of CNTF to block rod PR differentiation (Kirsch et al 1998, Schulz-Key et al 2002). CNTF expression is increased by rod degeneration in the *rd1* mouse retina (our results). The mechanisms for the side effects of CNTF administration should be determined before human clinical trials are considered for the amelioration of inherited retinal degeneration with CNTF (Bok et al 2002). While the possible benefits of a treatment using CNTF is matter of a debate, its delivery to the eye was successfully accomplished by mean of viral vectors, adenoviral (Cayouette & Gravel 1997) and adeno-associated viral vectors (Liang et al 2001) as well as alternatively using encapsulated cell line (Tao et al 2002). These studies have paved the way to the delivery of other therapeutic genes into the eye. The encapsulated cell-based delivery seems to be particularly appropriate in a disease that is not life-threatening since the capsules implanted into the eye can be removed by surgery. We plan to use this technology to test for the morphological and more importantly the functional benefit of RdCVF treatment.

Acknowledgements

The authors thank Aurélie Gluck, Juliette Ravey, Danièle Thiersé, Manuel Simmonutti, Valérie Forster, Noëlle Hanotteau, Emmanuelle Clérin, Pascal Oberlin and Georges Tarlet for excellent technical assistance. The authors thank Arno Dolemeyer (Novartis-Pharma, Basle, Switzerland), Christophe Grolleau (Roper Scientific, Evry, France) and Eugène Scherbeck (Couvoirs de l'Est, Wilgottheim, France) for their invaluable help. The authors thank Pierre Chambon for constant support. This work was financed by Novartis, Inserm, Ministère de la Recherche, the Association Française contre les Myopathies (AFM), the Fédération des Aveugles de France, Retina France, Foundation Fighting Blindness (USA), IPSEN Foundation, and the European Community (PRO-AGE-RET program).

References

Adler R, Hatlee M 1989 Plasticity and differentiation of embryonic retinal cells after terminal mitosis. Science 243:391–393

Bok D, Yasumura D, Matthes MT et al 2002 Effects of adeno-associated virus-vectored ciliary neurotrophic factor on retinal structure and function in mice with a P216L rds/peripherin mutation. Exp Eye Res 74:719–735

Bowes C, Li T, Danciger M, Baxter LC, Applebury ML, Farber DB 1990 Retinal degeneration in the rd mouse is caused by a defect in the beta subunit of rod cGMP-phosphodiesterase. Nature 347:677–680

Carter-Dawson LD, La Vail MM, Sidman RL 1978 Differential effect of the rd mutation on rods and cones in the mouse retina. Invest Ophthalmol Vis Sci 17:489–498

Cayouette M, Gravel C 1997 Adenovirus-mediated gene transfer of ciliary neurotrophic factor can prevent photoreceptor degeneration in the retinal degeneration (rd) mouse.
Hum Gene Ther 8:423–430

Chang GQ, Hao Y, Wong F 1993 Apoptosis: final common pathway of photoreceptor death in rd, rds and rhodopsin mutant mice. Neuron 11:595–605

Cideciyan AV, Hood DC, Huang Y et al 1998 Disease sequence from mutant rhodopsin allele to rod and cone photoreceptor degeneration in man. Proc Natl Acad Sci USA 95:7103–7108

Curcio CA, Millican CL, Allen KA, Kalina RE 1993 Ageing of the human photoreceptor mosaic: evidence for selective vulnerability of rods in central retina. Invest Ophthalmol Vis Sci 34:3278–3296

Curcio CA, Medeiros NE, Millican CL 1996 Photoreceptor loss in age-related macular degeneration. Invest Ophthalmol Vis Sci 37:1236–1249

Dowling JE 1987 The retina: an approachable part of the brain. Harvard Press, Cambridge MA

Faktorovich EG, Steinberg RH, Yasumura D, Matthes MT, LaVail MM 1990 Photoreceptor degeneration in inherited retinal dystrophy delayed by basic fibroblast growth factor. Nature 347:83–86

Fintz AC, Audo I, Hicks D, Mohand-Saïd S, Léveillard T, Sahel J 2003 Partial characterization of retinal derived cone neuroprotection in two culture models of photoreceptor degeneration. Invest Ophthalmol Vis Sci 44:818–825

Frasson M, Picaud S, Léveillard T et al 1999 Glial cell line-derived neurotrophic factor induces histologic and functional protection of rod photoreceptors in the rd/rd mouse. Invest Ophthalmol Vis Sci 40:2724–2734

Kajiwara K, Berson EL, Dryja TP 1994 Digenic retinitis pigmentosa due to mutations at the unlinked peripherin/RD and ROM I loci. Science 264:1604–1608

Kirsch M, Schulz-Key S, Wiese A, Fuhrmann S, Hofmann H 1998 Ciliary neurotrophic factor blocks rod photoreceptor differentiation from postmitotic precursor cells in vitro. Cell Tissue Res. 291:207–216

LaVail MM, Yasumura D, Matthes MT et al 1998 Protection of mouse photoreceptors by survival factors in retinal degenerations. Invest Ophthalmol Vis Sci 39:592–602

Liang FQ, Aleman TS, Dejneka NS et al 2001 Long-term protection of retinal structure but not function using RAAV.CNTF in animal models of retinitis pigmentosa. Mol Ther 4: 461–472

McCall MA, Gregg RG, Merriman K, Goto Y, Peachey NS, Stanford LR 1996 Morphological and physiological consequences of the selective elimination of rod photoreceptors in transgenic mice. Exp Eye Res 63:35–50

McLaughlin ME, Sandberg MA, Berson EL, Dryja TP 1993 Recessive mutations in the gene encoding the beta-subunit of rod phosphodiesterase in patients with retinitis pigmentosa. Nat Genet 4:130–134

Mohand-Said S, Hicks D, Simonutti M et al 1997 Photoreceptor transplants increase host cone survival in the retinal degeneration (rd) mouse. Ophthalmic Res 29:290–297

Mohand Saïd S, Deudon-Combe A, Hicks D et al 1998 Normal retina releases a diffusible factor stimulating cone survival in the retinal degeneration (rd) mouse. Proc Natl Acad Sci USA 95:8357–8362

Mohand Said S, Hicks D, Dreyfus H, Sahel J A 2000 Selective transplantation of rods delays cone loss in a retinitis pigmentosa model. Arch Ophthalmol 118:807–811

Noell WK 1980 Possible mechanisms of photoreceptor damage by light in mammalian eyes. Vision Res 20:1163–1171

Owsley C, Jackson GR, Cideciyan AV et al 2000 Psychophysical evidence for rod vulnerability in age-related macular degeneration. Invest Ophthalmol Vis Sci 41:267–273

Perry J, Du J, Kjeldbye H, Gouras P 1995 The effects of bFGF on RCS rat eyes. Curr Eye Res 14:585–592

Rosenfeld PJ, Cowley GS, McGee TL, Sandberg MA, Berson EL, Dryja TP 1992 A null mutation in the rhodopsin gene causes rod photoreceptor dysfunction and autosomal recessive retinitis pigmentosa. Nat Genet 1:209–213

Schulz-Key S, Hofmann HD, Beisenherz-Huss C, Barbisch C, Kirsch M 2002 Ciliary neurotrophic factor as a transient negative regulator of rod development in rat retina. Invest Ophthalmol Vis Sci 43:3099–3108

Sunness JS, Massof RW, Johnson MA, Bressler NM, Bressler SB, Fine SL 1989 Diminished foveal sensitivity may predict the development of advanced age-related macular degeneration. Ophthalmology 96:375–381

Tao W, Wen R, Goddard MB et al 2002 Encapsulated cell-based delivery of CNTF reduces photoreceptor degeneration in animal models of retinitis pigmentosa. Invest Ophthalmol Vis Sci 43:3292–3298

DISCUSSION

Hauswirth: What does the human version of this gene look like and where does it map relative to RD?

Sahel: We recently looked at the intron/exon sequence of the gene and there is a human counterpart as well as a rat counterpart. We are now defining the chromosomal location. We have plans to work on trying to find a mutation in cone degeneration, for example.

Nathans: Can you tell us something about what the protein looks like?

Sahel: The sequence doesn't look like a neurotrophin, for example. From looking at the sequence and doing some bioinformatics analysis, we haven't found an obvious function for this protein.

Ali: Don't you think that it is strange that it is localized in the outer segments, as opposed to the cell body? Can you speculate a little about this?

Sahel: I can't explain this. There was a nice paper by Cideciyan et al (1998) in which they showed by psychophysical studies that cones start to degenerate once you lose 75% of the outer segments of a photoreceptor. It is possible it is expressed at this level.

Hauswirth: Just because you detect the protein by antibody localization in the outer segments it doesn't mean that it is made in the rods. Have you looked at RNA *in situ* to see where it is made?

Sahel: No, but if we cut the retina in slices (and we get 99% photoreceptor purity in these preparations), the RNA expression is in the photoreceptor layer.

Travis: From the primary sequence does the protein look like a secreted protein?

Sahel: Yes.

LaVail: Did you say that conditioned medium from wild-type mice was effective whereas that from *rd1* mice was not?

Sahel: Yes.

LaVail: What was the age of the *rd* mouse?

Sahel: 5 weeks. At 8 days you get the effect but at 5 weeks there is no effect.

Hauswirth: We have tried many times to express and secrete a vectored gene product in a photoreceptor-specific way using a rod promoter and vector. We can't get anything secreted for rods. It may not apply to this situation, but I don't think a rod photoreceptor is a great source of secreted anything.

Farber: I don't agree. You can get things secreted from rods or cones.

Hauswirth: I stated our experience. There are a lot of endogenously secreted proteins in photoreceptors but we can't seem to engineer them get to secrete a vectored gene product under these circumstances.

Bok: It is a little weird to have a protein secreted by the outer segments. There is no secretory mechanism that I know of for proteins in the outer segment.

Dryja: Are you sure it is in the outer segments?

Sahel: It might be at the surface, or inside the outer segment.

Kaleko: Couldn't it be secreted from elsewhere in the photoreceptor?

Bok: IRBP is a good example. If you stain a retina for IRBP using the method that you have employed here, it would look as though the protein is within the outer segments. However, IRBP is extracellular and the zonulae adherens, which collectively form the outer limiting membrane, hold the IRBP in the subretinal space. It can't diffuse past that. If the protein that you describe is extracellular it must be bound to something, because it is a small protein of 25 kDa which could escape the confines of the outer limiting membrane. That is to say, it is small enough to get through the zonula adherens that form the outer limiting membrane. It is likely that it is bound to something in the extracellular space, which would make sense. Many neurotrophic factors bind there and it acts like a slow release capsule. This would make sense to me.

Travis: Is this protein glycosylated?

Sahel: We haven't looked yet. We are currently making transgenic animals over-expressing the factor in the RPE and we are crossing back to the *rd1* mouse to see whether we can rescue them.

Kaleko: In your assays can you distinguish a survival factor from a growth factor?

Sahel: No. We decided to call this a viability factor and not a survival factor. The reason for this is that cones are not directly undergoing any degeneration. They just need something to make them viable.

Kaleko: Is that similar to GDNF and CNTF? Do they really cause improved viability?

LaVail: That's a semantic distinction. The term 'survival' factor is used in neurotrophic factor parlance as distinct from a differentiation factor.

Kaleko: Do any of them cause cell proliferation in the retina?

LaVail: They do, under certain circumstances.

Kaleko: I was thinking more along the lines of RPE or photoreceptors. Can these factors reverse a disease phenotype or are they more likely to slow the disease course?

LaVail: People haven't studied this as thoroughly as we would like. There have not been many cases where we see the sort of overabundance that might result from a reinstitution of proliferation at a later stage.

Travis: In addition to this very interesting protein did you find any growth factors such as CNTF in your clones?

Sahel: CNTF is not one of the 24. In the differential expression in the chips, there is a 10-fold increase in CNTF mRNA expression after rod degeneration rather than the loss of expression. One of the criteria we used for identifying the clone was loss of expression after the rods have degenerated. CNTF did not fit this criterion. Our hypothesis was that there was loss of something.

LaVail: CNTF is up-regulated in almost all kinds of injury. Even normal cell death can cause its expression in surrounding cells. We should recognize that in some species neural cells begin to proliferate in response to injury and loss of cells, such as retinal detachment. It is not out of the question that some retinal cells might proliferate.

Swaroop: When are the rods completely degenerated? Is this factor still detectable after all rods have died?

Sahel: Yes, which is why I discussed the possibility that it is a photoreceptor factor. In the rd1 mouse there are still a few rods remaining at the periphery.

Dryja: I remember hearing of an ophthalmologist who had decided to treat RP patients by laser. He/she photocoagulated the boundary between the degenerated retina and the remaining viable retina. If your data are correct about these survival factors, then killing off even more photoreceptors can't be of any benefit.

Sahel: CNTF might be released by the cells.

LaVail: And also FGF. In almost every case where there is injury, FGF is up-regulated (reviewed by Gao & Hollyfield 1996). It is the prime candidate for the molecule that is effecting neurotrophic factor rescue. In other words, any way you can up-regulate FGF will protect photoreceptor cells. With regard to Dr Dryja's comment about laser treatment, it has been shown by Chu et al (1998) that in RCS rats with inherited retinal dystrophy, laser photocoagulation leads to FGF2 up-regulation in surrounding regions at precisely the areas where photoreceptors are transiently protected from degeneration.

Dryja: Are you advocating this as a possible therapy?

LaVail: No, I'm just saying that there is a reason why it might work.

Sieving: The laser work in human retinitis pigmentosa patients was being done at Ohio State University by Dr Davidorf nearly two decades ago, and there was no positive benefit reported.

Dryja: I think a pharmaceutical high-throughput screen for a small molecule that mimics the function of the survival factor that could be systemically delivered would probably be quite valuable.

References

Chu Y, Humphrey MF, Alder VV, Constable IJ 1998 Immunocytochemical localization of basic fibroblast growth factor and glial fibrillary acidic protein after laser photocoagulation in the Royal College of Surgeons rat. Aust NZ J Ophthalmol 26:87–96

Cideciyan AV, Hood DC, Huang Y et al 1998 Disease sequence from mutant rhodopsin allele to rod and cone photoreceptor degeneration in man. Proc Natl Acad Sci USA 95:7103–7108

Gao H, Hollyfield JG 1996 Basic fibroblast growth factor: increased gene expression in inherited and light-induced photoreceptor degeneration. Exp Eye Res 62:181–189

Studies on retinal and retinal pigment epithelial gene expression

Itay Chowers*[1], Noriko Esumi*[1], Peter Campochiaro*† and Donald J. Zack*†‡§[2]

*Guerrieri Center for Genetic Engineering and Molecular Ophthalmology at the Wilmer Eye Institute, the Departments of †Neuroscience, and ‡Molecular Biology and Genetics, and §McKusick-Nathans Institute of Genetic Medicine, The Johns Hopkins University School of Medicine, Baltimore, Maryland, MD 21287, USA

Abstract. The 'completion' of the murine and human genomes and creation of high-density expressed sequence tag (EST) databases from multiple tissues and multiple species, coupled with the development of high-throughput expression profiling approaches such as microarrays and Serial Analysis of Gene Expression (SAGE), is making possible the in-depth analysis of gene expression patterns in health and disease to an extent that was not previously possible. Such new information is providing insight into normal function, and into how normal function is altered in disease. Efforts have begun, and are accelerating, in the application of expression profiling to the study of the retina and retinal pigment epithelium (RPE). In this chapter we will review progress in this area. We will also discuss technical issues that make expression studies of the RPE particularly challenging, and share our experience in methodological approaches to overcome these challenges.

2004 Genetics to gene therapy of retinal dystrophies. Wiley, Chichester (Novartis Foundation Symposium 248) p 131–146

Approaches for high-throughput gene expression profiling

Serial Analysis of Gene Expression (SAGE), one of the major methods for expression profiling, is based on the sequencing of short DNA fragments, known as 'tags', from the 3′ end region of a representative set of the mRNA transcripts from a cell or tissue sample (Velculescu et al 1995). Since SAGE can theoretically identify novel as well as known genes, as long as sufficient genomic data are available, it is a powerful method for gene discovery as well as gene

---

[1]These authors have contributed equally to the chapter.
[2]This paper was presented at the symposium by Donald J. Zack, to whom correspondence should be addressed.

expression studies. Unfortunately, although the situation is changing with the decreasing cost of DNA sequencing, the application of SAGE to multiple samples can be both cost and labour intensive. Thus SAGE is useful for studying a few samples in depth, but has limitations for the study of many samples, such as comparing a large group of normal and disease specimens.

DNA microarrays, which can be either cDNA or oligonucleotide based, allow the simultaneous expression profiling of thousands of genes, and multiple replicates can be easily performed. Compared to SAGE, microarrays have the advantage that, once established, they require limited labour and are relatively rapid and cost effective. However, a major limitation of microarray technology compared to SAGE is that gene expression levels can be assayed only for genes that are represented on the microarray slide (if a gene is not on the array you cannot 'see' it), and the expression levels of different genes cannot be directly compared.

Bioinformatic analysis of large expressed sequence tag (EST) databases provides another approach for gene discovery and expression profiling (Bernal-Mizrachi et al 2003, Strausberg et al 2003). As with SAGE it theoretically allows comparison of expression levels across genes, but it is also subject to various artefacts and sources of bias (Audic & Claverie 1997). Analysis of multiple samples and incorporation of replicates in the experimental design can be difficult and costly.

Although one would hope that SAGE, microarray, and EST-based approaches would provide similar expression profiles, the degree of overlap between these different approaches has not been carefully evaluated. In a preliminary effort to address this issue, we recently compared the results of four studies focused on identifying novel retina-enriched genes (two SAGE studies, one EST data mining and one microarray study) (Stohr et al 2000, Blackshaw et al 2001, Sharon et al 2002, Chowers et al 2003a). While the microarray and EST data mining studies shared the identification of several novel retina-enriched genes, the overlap with the SAGE results was limited. This finding emphasizes the need for caution in interpreting expression studies, and the need for further work to explore the reasons why these different approaches can yield varying results.

The normal and diseased retina is an attractive target for gene expression profiling studies for a number of reasons (Swaroop & Zack 2002). However, it should also be noted that there are a number of challenges. Among these are that commercial arrays tend to have limited representation of retina-specific genes, the many different cell types within the retina can mask expression changes in small subpopulations and can make analysis difficult, and acquiring good quality human donor retina RNA can be difficult, especially for disease specimens. Although the latter problem can be partially addressed by using animal tissue, there are significant differences between primate and non-primate retinas, and there are significant limitations in many of the animal models of human retinal disease. Below we provide examples of how various investigators have dealt with these issues.

Gene expression profiling in exploration of retinal development

Retinal development involves the differentiation of uncommitted neuroblasts into multiple cell types in a complex anatomical organization, a process requiring carefully choreographed and timed changes in gene expression. As a step in the further characterization of these gene expression changes, Mu et al (2001) constructed a cDNA library from embryonic day 14.5 (E14.5) mouse retina, a stage at which the retina is composed mostly of uncommitted neuroblasts but with some newly formed differentiated neurons. Annotation of about 9000 sequences from the library enabled estimation of the retinal gene expression profile at that embryonic stage, thereby providing a list of candidate genes that may play a role in retina development (Mu et al 2001). The authors then constructed a microarray containing 864 cDNAs selected from the retina library, and compared retinal expression level between wild-type and $Brn3b^{-/-}$ mice. The analysis identified Gap43, a molecule implicated in retinal ganglion cell axon growth and pathfinding, as a potential target of the $Brn3b$ transcription factor.

Diaz et al (2003) recently reported a statistically sophisticated study of murine retinal development utilizing the Riken 19K cDNA array. They analysed spatial expression patterns across E14.5 retinas by separating hybridizing nasal, temporal, dorsal and ventral regions. Using a clustering approach, they identified groups of genes with distinct spatial expression patterns. Many of the gradient patterns identified were confirmed by *in situ* hybridization.

In order to study the transcriptional network of the photoreceptor homeobox transcription factor CRX, Livesey et al (2000) used a microarray that included 960 cDNAs from a mouse retina cDNA library. The authors identified 16 genes that showed differential expression pattern in normal postnatal day 10 $Crx^{+/+}$ mice retina as compared to the retina of $Crx^{-/-}$ mice. Fifteen of the 16 genes were more highly expressed in the $Crx^{+/+}$ mouse, and the majority of these genes were either photoreceptor specific or corresponded to ESTs from unknown genes. By analysing the promoter sequences of some of the differentially expressed genes the authors identified a novel motif that is similar to the standard homodomain transcription factor binding site. It is noteworthy that the authors used a polymerase chain reaction (PCR) amplification technique to increase the amount of cDNA probe, and presented data suggesting reasonable reproducibility of the amplification procedure. This is significant because many types of potentially interesting retinal microarray studies, especially those utilizing developing retinal and single cell preparations, are limited by the amount of available RNA.

Gene expression profiling in normal adult retina

Identification of tissue-specific genes continues to be a major focus in retina research, both because of the crucial function such genes play in the normal retina and because of the association of mutation in many of these genes with retinal diseases (Blackshaw et al 2001). Several bioinformatic studies based on EST data mining successfully identified a number of retina-enriched genes (Shimizu-Matsumoto et al 1997, Malone et al 1999, Bortoluzzi et al 2000, Stohr et al 2000, Wistow et al 2002, Sinha et al 2000). In an alternative approach, Sharon and colleagues utilized SAGE to study two human retina samples. By comparing a SAGE library constructed from the macula to a library constructed from peripheral retina they identified a number of genes that appear to be expressed predominantly in the macula or the periphery. Similarly, by comparing a SAGE library from a 44-year-old donor to a library from an 88-year-old donor the authors identified candidate age-associated genes. The authors then identified retina-enriched genes by comparing the retina SAGE libraries to SAGE libraries from other tissues. This study demonstrated the power of SAGE in detecting expression profiles of novel genes. However, it also demonstrated a limitation of SAGE, in that the expression differences observed between the samples could be related to multiple technical and biological parameters unrelated to those studied by the authors. This could have been determined by analysing replicate samples, but this was not done because of the time and expense of SAGE replicate studies. In a related set of studies, Blackshaw and colleagues applied SAGE to the mouse retina (Blackshaw et al 2001). By analysing data from SAGE libraries of different retina developmental stages, brain, dissected photoreceptor layer, and mutant retinas, the authors identified 264 uncharacterized photoreceptor enriched-genes.

Because commercial arrays do not include many genes of interest to retina biologists, a number of groups have generated custom arrays enriched for genes of importance to the retina (Livesey et al 2000, Mu et al 2001, Farjo et al 2002, Yu et al 2002). We have constructed both murine and human custom retina cDNA microarrays (Hackam et al 2003, Chowers et al 2003a). The human array contains PCR products representing approximately 10 000 genes, with 67% corresponding to known genes and 33% corresponding to ESTs. The sequences were selected in an effort to reflect the predicted human retina gene expression profile, based on EST databases, and also to include genes that we thought might be involved in pathways related to retinal development, function, and disease. By comparing human retina, liver, and cerebral cortex, we have identified 186 genes with a previously uncharacterized retina-enriched expression pattern (Chowers et al 2003a).

Another area of interest in which high-throughput profiling technology is having an effect is in the understanding of gene expression variation between normal individuals. Expression variation across different genes, tissues, and individuals, in addition to structural gene variation, is a major source of phenotypic diversity in both normal and diseased tissue (Enard et al 2002, Oleksiak et al 2002, Cheung et al 2003, Whitney et al 2003). In the eye, analysis of micro-SAGE retina libraries has suggested that there is significant expression variation in the retina, even between different inbred mice (Blackshaw et al 2003). We have used our human microarray to further explore this issue (Chowers et al 2003b). Duplicate microarrays were performed on 33 adult human retina samples, from 19 individuals. The resulting data were analysed in order to quantify the degree of expression variation between individuals and between genes, and also assessed to determine the association of variation with identifiable biological factors as gene function, age and gender. We found the mean expression ratio standard deviation between individuals to be relatively small (0.15 ± 0.8, $\log_2$), but there was a wide range (from 0.09–0.99), meaning that the expression of some genes varied widely between different normal individuals. This information will hopefully be useful in interpreting ongoing studies by a number of groups that are aimed at determining expression differences between normal and diseased retinas (see below).

Microarrays are also being used to characterize gene expression changes associated with aging, with studies ranging from *Drosophila* to human (Helmberg 2001, Jin et al 2001). Several efforts have focused on the retina. Bernstein and colleagues used cDNA membrane macroarrays to identify age-associated gene expression in the human retina, and reported down-regulation of the heat shock cognate protein 70 in primate retinas (Bernstein et al 2000). Yoshida and colleagues applied a commercial cDNA microarray with 2400 genes to assess age-related expression changes in five human retinas, two of them from young donors (age 13 and 14 years) and three from older donors (age 62–72 years) (Yoshida et al 2002). Twenty-four genes were identified as having a differential expression pattern across the age groups. As part of the study cited above, we have recently identified several additional age-associated genes in the human retina by analysing expression profiles of 33 retinas from 19 donors aged 29–90 years (Chowers et al 2003b).

Gene expression profiling in retinal disease

Several recent studies highlight the ability of high throughput expression studies to advance our understanding of retinal disease. Such studies have the potential both to help identify the primarily mutated gene in genetic disorders and to identify downstream pathways involved in disease pathogenesis. Blackshaw et al (2001) analysed several murine retinal SAGE libraries and identified 87 novel retina-enriched genes that map to 37 loci linked to retinal disease. Aided by this data, Bowne et al (2002) identified a mutation in the gene encoding inosine monophosphate dehydrogenase type 1 (IMPDH1) in several families with autosomal dominant retinitis pigmentosa (RP10 locus, at 7q31.1). In a striking example of the increasing convergence of expression and genetic technologies, at about the same time, and independently, Kennan et al (2002), on the basis of hints from microarray data, also identified mutations in IMPDH1 as responsible for RP10. Their approach involved the use of Affymetrix GeneChips to compare expression in retinas from normal vs. rhodopsin knockout mice. In their analysis, IMPDH1 showed an average decrease in expression of 5.8-fold in the knockout mice.

Altered transcription levels of retinal genes can also reflect secondary changes in the degenerating retina. For example, Jones and colleagues reported that in the *rd1* retina there is up-regulation of the stress-response related gene αB-crystallin, mostly in glial cells (Jones et al 1998), and up-regulation of the *nm23* gene, predominantly in ganglion cell layer cells (Jones et al 2000a). The authors speculate that increased αB-crystallin and nm23-M2 expression in the retina may be part of the response to the degenerative process. In contrast, in the Royal College of Surgeons (RCS) rat degeneration model a two-dimensional polyacrylamide gel electrophoresis based study demonstrated down-regulation of αA-crystallin and rhodopsin kinase (Maeda et al 1999). In another study, Jones et al (2000b) used membrane cDNA macroarray to compare the expression profile of 205 apoptosis-related genes in human tissue, comparing retinitis pigmentosa (RP) and normal donor retinas. Among the few genes that showed significant expression differences were c-Jun, PIG7 and secreted Frizzled-related protein-2 (sFRP-2). The presence of increased levels of sFRP-2 in RP retinas was confirmed by Northern blot and immunohistochemical analysis. Identification of such deviations from normal retinal gene expression patterns can serve to uncover the mechanisms that underline photoreceptor and retinal pigment epithelium death in retinal degenerations, as well as the impact of the primary mutation and the degeneration process on other parts of the retina.

Early events in the retinal response to light toxicity have been studied using microarray technology. Choi et al (2001) identified groups of differentially expressed genes in arrestin/rhodopsin kinase-deficient mouse retinas after exposure to light. Some of these gene expression changes occurred long before degenerative morphological changes could be identified. Microarrays have also been used to identify differentially expressed genes in a mouse model of Norrie disease, and to determine the time point where these changes occur (Lenzner et al 2002). Additional microarray studies have assessed retinal gene expression in animal models of retinal damage from ischaemia reperfusion (Yoshimura et al 2003), diabetic retinopathy (Cho et al 2002), retinal laser photocoagulation (Wilson et al 2003), and photopic retinal injury (Chen et al 2003). These studies have identified candidate genes for involvement in retinal damage in these diseases. Microarray analysis was also recently applied to characterize gene expression changes in Weri-Rb-1 retinoblastoma cells induced by retinoic acid, with the predominant finding being a tendency to increase expression of cone-associated genes (Li et al 2003).

Microarray are also beginning to be applied to the study of glaucoma. Laabich and colleague used the Clontech rat Atlas expression array, a membrane cDNA macroarray, to study rat retinal gene expression changes induced by N-methyl-d-aspartate (NMDA) (Laabich et al 2001). They identified five genes (*FasL*, *GADD45*, *GADD153*, *Nur77*, and *TNF-R1*) that may be related to NMDA induced cell death, based on the finding of increased expression in the range of 2–7 fold two hours after intravitreal NMDA injection. These results were confirmed by reverse transcription PCR (RT-PCR). They also demonstrated that increased expression of these genes could be blocked by AIP (myristoylated autocamtide-2-related inhibitory peptide), which inhibits calcium/calmodulin-dependent protein kinase II. Using microarrays, Hernandez and colleagues identified multiple genes that are differentially expressed between cultured astrocytes from glaucomatous and normal optic nerves (Hernandez et al 2002). Gonzalez et al (2000) used macroarrays to study the effect of elevated intraocular pressure on human trabecular meshwork gene expression.

Retinal pigment epithelium gene expression

Studies of retinal pigment epithelium (RPE) gene expression share many of the challenges discussed above for retinal expression studies. In addition, there are challenges unique to the RPE. These include the greater difficulty in isolating 'pure' RPE (this is especially true for murine RPE) and the lower abundance of RPE tissue compared to retina, and hence the lower abundance of RPE RNA. In our experience, we obtain less than 10% as much RPE RNA as retinal RNA in from a typical human donor eye. (Advantage — single, largely homogeneous cell type.) An additional technical challenge is posed by the presence of melanin pigment within the RPE, which can interfere with the purification and subsequent enzymatic analysis of RPE RNA.

It has long been recognized that melanin co-purifies with nucleic acids, both DNA and RNA, from pigmented cells such as melanocytes and melanoma cells, and can inhibit PCR and RT-PCR (Yoshii et al 1992, 1993, Giambernardi et al 1998, Price & Linge 1999). Of more direct relevance to this discussion, the same problem has also been encountered when RT-PCR was performed with RNA from several pigmented ocular tissues such as iris, ciliary body, and choroid/RPE (Wang et al 2001, Kyveris et al 2002). Melanin has been shown to be a potent inhibitor of thermostable DNA polymerase through its direct binding to the enzyme (Eckhart et al 2000). Melanin blocks PCR at concentrations below 200 ng/ml, while 100 μg/ml melanin is required to inhibit the RT step (Eckhart et al 2000). Therefore, it has been suggested that the inhibition of RT-PCR by melanin contaminated in RNA preparations occurs mainly at the PCR step, and the melanin effect is marginal at the RT step (Giambernardi et al 1998, Eckhart et al 2000). However, during the construction of human RPE cDNA libraries for yeast one/two-hybrid systems, we observed that RNA extracted from human RPE was still dark brown even after double selection of poly(A)$^+$ RNA by oligo(dT) cellulose, and the RT extension with this coloured poly(A)$^+$ RNA was substantially inhibited. These observations led us to search for strategies to overcome the inhibition of RT for constructing well-represented RPE cDNA libraries.

We initially used TRIzol reagent (Invitrogen, Carlsbad, California, USA) to extract total RNA from human RPE, followed by MessageMaker oligo(dT) cellulose (Invitrogen) for double selection of poly(A)$^+$ RNA. The TRIzol reagent involves a modification of the single-step RNA isolation method by acid guanidinium isothiocyanate (GITC)-phenol-chloroform extraction (Chomczynski & Sacchi 1987). We observed that the total RNA isolated by TRIzol from human RPE was dark brown, and that double selection by oligo(dT) cellulose did not decrease the coloration. First-strand cDNA synthesis using Stratascript reverse transcriptase (Stratagene, La Jolla, California, USA) with this brown poly(A)$^+$ RNA failed to produce detectable cDNA. Therefore, we performed several pilot experiments to look for strategies to circumvent this problem. Because of its greater availability compared to human, we used bovine RPE for these pilot studies. We first tested the same methods that had been used for human RPE to isolate poly(A)$^+$ RNA from bovine RPE and mouse liver in parallel and performed the RT reaction with either oligo(dT) primer or random 15-mer. Both total and poly(A)$^+$ RNA from bovine RPE were pigmented (brown), although they were lighter than those from human RPE. In contrast, both RNAs from mouse liver were completely pigment-free (white). Analysis by formaldehyde agarose gel electrophoresis showed that the quality of the total RNAs from two tissues was equivalent (data not shown). First-strand cDNA synthesis by either oligo(dT) or random priming, monitored by incorporation of [α^{32}P]dATP, was substantially more efficient with poly(A)$^+$ RNA from mouse liver than that from bovine RPE (Fig. 1). The RT extension by oligo(dT) primer was significantly longer than that by random primer with mouse liver RNA. First-strand cDNA generated by random priming was undetectable with bovine RPE RNA (Fig. 1). The same methods were also applied for human retina, and the results were very similar to those obtained with mouse liver (data not shown). These results suggested that the primary cause of the RT inhibition was contamination of melanin pigment.

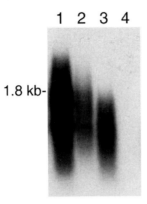

FIG. 1.   Comparison of first-strand cDNA synthesis with poly(A)$^+$ RNA from bovine RPE
and mouse liver. Total RNA was extracted by TRIzol followed by poly(A)$^+$ RNA isolation
using double selection with MessageMaker oligo(dT) cellulose. The RT reaction was
performed by either oligo(dT) (lanes 1 and 3) or random priming (lanes 2 and 4) using
Stratascript reverse transcriptase. The first-strand cDNAs were labelled by incorporation of
$[\alpha^{32}P]dATP$ and analysed on a 1% alkaline agarose gel. The gel was dried with Whatman 3MM
papers and exposed to X-ray film. The first-strand cDNA synthesis was significantly more
efficient with poly(A)$^+$ RNA from mouse liver (lanes 1 and 2) than that from bovine RPE
(lanes 3 and 4).

We then compared three different methods for poly(A)$^+$ RNA isolation from
bovine RPE: (1) total RNA extraction by TRIzol followed by double selection of
poly(A)$^+$ RNA using MessageMaker (TRIzol method), (2) total RNA extraction
by RNeasy (Qiagen, Valencia, California, USA) followed by double selection of
poly(A)$^+$ RNA using Oligotex (Qiagen) (RNeasy/Oligotex method), and (3)
direct poly(A)$^+$ RNA isolation using the Oligotex direct mRNA protocol
followed by an additional single selection using Oligotex (Oligotex method).
RNeasy is a silica-gel-based membrane method in a micro-spin column format
devoid of phenol/chloroform extraction and alcohol precipitation. Oligotex
consists of polystyrene-latex particles to which $dC_{10}T_{30}$ oligonucleotides are
covalently linked. Bovine eyes were obtained from a local slaughterhouse, and
within two hours from the time of death the RPE layer of each eye was dissected
into 1 ml ice-cold PBS in a microcentrifuge tube. The RPE cell pellets were
collected by centrifugation at 6000 rpm for 5 min at 4 °C, flash frozen in liquid
nitrogen, and stored at −70 °C until extraction. The RPE pellet from one eye was
processed using 1 ml TRIzol, an RNeasy mini column, or an Oligotex mini-prep
according to the respective manufacturer's instructions. A QIAshredder spin
column (Qiagen) was used to homogenize each RPE pellet when RNeasy or
Oligotex was employed. While the poly(A)$^+$ RNA isolated by the TRIzol

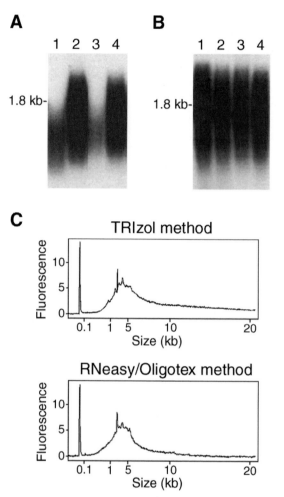

FIG. 2.   Comparison of poly(A)$^+$ RNA isolated by different methods from bovine RPE. Poly(A)$^+$ RNA was purified by either (1) total RNA extraction by TRIzol followed by double selection of poly(A)$^+$ RNA using MessageMaker (TRIzol method) or (2) total RNA extraction by RNeasy followed by double selection of poly(A)$^+$ RNA using Oligotex (RNeasy/Oligotex method). (A) The RT reaction and the sample processing were same as described in Fig. 1. The first-strand cDNA synthesis was clearly more efficient with poly(A)$^+$ RNA isolated by the RNeasy/Oligotex method (lanes 2 and 4) than that by the TRIzol method (lanes 1 and 3). Priming: oligo(dT) primer (lanes 1 and 2) or random primer (lanes 3 and 4). (B) The RT reaction was performed using SuperScript II (200 units/μl) instead of Stratascript (50 units/μl). Samples (lanes 1–4) are in the same order as in A. The first-strand cDNA synthesis was similar in all samples. (C) There was no significant difference in the quality of the poly(A)$^+$ RNA prepared by the two methods, as assessed using an Agilent Bioanalyzer. The ribosomal RNA contamination of the poly(A)$^+$ RNA was similar, approximately 1.5% after double selection by both methods.

method was coloured brown, both the RNeasy/Oligotex and Oligotex methods successfully removed melanin pigment. First-strand cDNA synthesis by either oligo(dT) or random priming was clearly more efficient with poly(A)$^+$ RNA isolated by the RNeasy/Oligotex method than that by the TRIzol method (Fig. 2A).

Since melanin binds directly to thermostable DNA polymerase (Eckhart et al 2000), we hypothesized that melanin might also bind to reverse transcriptase, and the RT inhibition by melanin might be due to reduction of the free active enzyme concentration. Therefore, we investigated the effect of increasing the total units of reverse transcriptase in the reaction mix. Indeed, the RT inhibition by melanin was overcome by this strategy. Similar first-strand cDNA syntheses were observed with pigmented and non-pigmented poly(A)$^+$ RNA in both oligo(dT) and random priming using a threefold larger amount of Stratascript (50 units/μl) (data not shown) or SuperScript II reverse transcriptase (200 units/μl, Invitrogen) (Fig. 2B).

The quality of poly(A)$^+$ RNA was analysed using the RNA 6000 Nano Assay with an Agilent Bioanalyzer (Agilent Technologies, Palo Alto, California, USA). This assay is based on an electrophoretic separation of a small volume of samples through micro-channels filled with polymer and fluorescence dye by strong electrokinetic driving forces. This analysis showed no significant difference in the quality of poly(A)$^+$ RNA between the TRIzol and RNeasy/Oligotex method (Fig. 2C). The ribosomal RNA contamination in the poly(A)$^+$ RNA was similarly $\sim$8% after a single selection with either MessageMaker or Oligotex (data not shown) and $\sim$1.5% after double selection by either method (Fig. 2C). The Oligotex method yielded equivalent results in the quality of poly(A)$^+$ RNA and first-strand cDNA synthesis to those by the RNeasy/Oligotex method (data not shown).

Our findings are consistent with previously reported strategies for RT-PCR of pigmented ocular tissues (Wirtz et al 1998, Wang et al 2001, Kyveris et al 2002). A major difference between the RNeasy/Oligotex and Oligotex methods and the widely used GITC-based single-step RNA isolation method (TRIzol) is that they do not utilize phenol/chloroform extraction and alcohol precipitation. Under normal physiological conditions, melanin is localized in membrane-enveloped vesicles and thus is prevented from interfering with cellular components outside the melanosome. In the process of RNA extraction, all cellular components are released into the same solution upon lysis of cellular and vesicular membranes. We suggest that the steps of alcohol precipitation and subsequent resuspension of RNA in a small volume may bring these cellular components close together and allow melanin to interact with RNA. Our results also demonstrate that the inhibitory effect of melanin can be overcome by increasing the total units of reverse transcriptase. If the co-purification of melanin with RNA is inevitable, this can be a simple strategy to overcome the problem. Although we have not yet performed direct comparison tests, we anticipate that the results we have presented here, which were generated in the context of cDNA library construction, will also be useful for increasing the success of microarray- and SAGE-based studies of RPE gene expression.

Acknowledgements

The authors thank Dr Anthony Lanahan for suggestions in regard to methods of RNA isolation and construction of cDNA libraries. This work was supported by funds from the NIH, Foundation Fighting Blindness, Macula Vision Foundation, and Research to Prevent Blindness, Inc., and generous gifts from Mr and Mrs Robert and Clarice Smith, and Mr and Mrs Marshall and Stevie Wishnack. P.A.C. is the George S. and Dolores Dore Eccles Professor of Ophthalmology and D.J.Z. is the Guerrieri Professor of Genetic Engineering and Molecular Ophthalmology.

References

Audic S, Claverie JM 1997 The significance of digital gene expression profiles. Genome Res 7:986–995

Bernal-Mizrachi EC, Cras-Meneur, Ohsugi M, Permutt MA 2003 Gene expression profiling in islet biology and diabetes research. Diabetes Metab Res Rev 19:32–42.

Bernstein SL, Liu AM, Hansen BC, Somiari RI 2000 Heat shock cognate-70 gene expression declines during normal aging of the primate retina. Invest Ophthalmol Vis Sci 41:2857–2862

Blackshaw S, Fraioli RE, Furukawa T, Cepko CL 2001 Comprehensive analysis of photoreceptor gene expression and the identification of candidate retinal disease genes. Cell 107:579–589

Blackshaw S, Kuo WP, Park PJ et al 2003 MicroSAGE is highly representative and reproducible but reveals major differences in gene expression among samples obtained from similar tissues. Genome Biol 4:R17

Bortoluzzi S, d'Alessi F, Danieli GA 2000 A novel resource for the study of genes expressed in the adult human retina. Invest Ophthalmol Vis Sci 41:3305–3308

Bowne SJ, Sullivan LS, Blanton SH et al 2002 Mutations in the inosine monophosphate dehydrogenase 1 gene (IMPDH1) cause the RP10 form of autosomal dominant retinitis pigmentosa. Hum Mol Genet 11:559–568

Chen L, Dentchev T, Wong R et al 2003 Increased expression of ceruloplasmin in the retina following photic injury. Mol Vis 9:151–158

Cheung VG, Conlin LK, Weber TM et al 2003 Natural variation in human gene expression assessed in lymphoblastoid cells. Nat Genet 33:422–425

Cho GJ, Ryu S, Kim YH et al 2002 Upregulation of glucose-dependent insulinotropic polypeptide and its receptor in the retina of streptozotocin-induced diabetic rats. Curr Eye Res 25:381–388

Choi S, Hao W, Chen CK, Simon MI 2001 Gene expression profiles of light-induced apoptosis in arrestin/rhodopsin kinase-deficient mouse retinas. Proc Natl Acad Sci USA 98:13096–13101

Chomczynski P, Sacchi N 1987 Single-step method of RNA isolation by acid guanidinium thiocyanate-phenol-chloroform extraction. Anal Biochem 162:156–159

Chowers I, Gunatilaka TL, Farkas RH et al 2003a Identification of novel genes preferentially expressed in the retina using a custom human retina cDNA microarray. Invest Ophthalmol Vis Sci 44:3732–3741

Chowers I, Liu D, Farkas RH et al 2003b Gene expression variation in the adult human retina. Hum Mol Genet, in press

Diaz E, Yang YH, Ferreira T et al 2003 Analysis of gene expression in the developing mouse retina. Proc Natl Acad Sci USA 100:5491–5496

Eckhart L, Bach J, Ban J, Tschachler E 2000 Melanin binds reversibly to thermostable DNA polymerase and inhibits its activity. Biochem Biophys Res Commun 271:726–730

Enard W, Khaitovich P, Klose J et al 2002 Intra- and interspecific variation in primate gene expression patterns. Science 296:340–343

Farjo R, Yu J, Othman MI et al 2002 Mouse eye gene microarrays for investigating ocular development and disease. Vision Res 42:463–470

Giambernardi TA, Rodeck U, Klebe RJ 1998 Bovine serum albumin reverses inhibition of RT-PCR by melanin. Biotechniques 25:564–566

Gonzalez P, Epstein DL, Barras T 2000 Genes upregulated in the human trabecular meshwork in response to elevated intraocular pressure. Invest Ophthalmol Vis Sci 41:352–361

Hackam AS, Strom R, Liu D et al 2003 Identification of gene expression changes implicated in progression of retina degeneration in the rd1 mouse using a custom retina microarray. Submitted.

Helmberg A 2001 DNA-microarrays: novel techniques to study aging and guide gerontologic medicine. Exp Gerontol 36:1189–1198

Hernandez MR, Agapova OA, Yang P, Salvador-Silva M, Ricard CS, Aoi S 2002 Differential gene expression in astrocytes from human normal and glaucomatous optic nerve head analyzed by cDNA microarray. Glia 38:45–64

Jin W, Riley RM, Wolfinger RD, White KP, Passader-Gurgel G, Gibson G 2001 The contributions of sex, genotype and age to transcriptional variance in *Drosophila melanogaster*. Nat Genet 29:389–395

Jones SE, Jomary C, Grist J, Thomas MR, Neal MJ 1998 Expression of alphaB-crystallin in a mouse model of inherited retinal degeneration. Neuroreport 9:4161–4165

Jones SE, Jomary C, Grist J, Stewart HJ, Neal MJ 2000a Identification by array screening of altered nm23-M2/PuF mRNA expression in mouse retinal degeneration. Mol Cell Biol Res Commun 4:20–25

Jones SE, Jomary C, Grist J, Stewart HJ, Neal MJ 2000b Altered expression of secreted frizzled-related protein-2 in retinitis pigmentosa retinas. Invest Ophthalmol Vis Sci 41:1297–1301

Kennan A, Aherne A, Palfi A 2002 Identification of an IMPDH1 mutation in autosomal dominant retinitis pigmentosa (RP10) revealed following comparative microarray analysis of transcripts derived from retinas of wild-type and Rho(−/−) mice. Hum Mol Genet 11:547–557

Kyveris A, Maruscak E, Senchyra M 2002 Optimization of RNA isolation from human ocular tissues and analysis of prostanoid receptor mRNA expression using RT-PCR. Mol Vis 8:51–58

Laabich A, Li G, Cooper NG 2001 Characterization of apoptosis-genes associated with NMDA mediated cell death in the adult rat retina. Brain Res Mol Brain Res 91:34–42

Lenzner S, Prietz S, Feil S, Nuber VA, Ropers HH, Berger W 2002 Global gene expression analysis in a mouse model for Norrie disease: late involvement of photoreceptor cells. Invest Ophthalmol Vis Sci 43:2825–2833

Li A, Zhu X, Brown B, Croft CM 2003 Gene expression networks underlying retinoic acid-induced differentiation of human retinoblastoma cells. Invest Ophthalmol Vis Sci 44:996–1007

Livesey FJ, Furukawa T, Steffen MA, Church GM, Cepko CL 2000 Microarray analysis of the transcriptional network controlled by the photoreceptor homeobox gene Crx. Curr Biol 10:301–310

Maeda A, Ohguro H, Maeda T, Nakagawa T, Kuroki Y 1999 Low expression of alphaA-crystallins and rhodopsin kinase of photoreceptors in retinal dystrophy rat. Invest Ophthalmol Vis Sci 40:2788–2794

Malone K, Sohocki MM, Sullivan LS, Daiger SP 1999 Identifying and mapping novel retinal-expressed ESTs from humans. Mol Vis 5:5

Mu X, Zhao S, Pershad R et al 2001 Gene expression in the developing mouse retina by EST sequencing and microarray analysis. Nucleic Acids Res 29:4983–4993

Oleksiak MF, Churchill GA, Crawford DL 2002 Variation in gene expression within and among natural populations. Nat Genet 32:261–266

Price K, Linge C 1999 The presence of melanin in genomic DNA isolated from pigmented cell lines interferes with successful polymerase chain reaction: a solution. Melanoma Res 9:5–9

Sharon D, Blackshaw S, Cepko CL, Dryja TP 2002 Profile of the genes expressed in the human peripheral retina, macula, and retinal pigment epithelium determined through serial analysis of gene expression (SAGE). Proc Natl Acad Sci USA 99:315–320

Shimizu-Matsumoto A, Adachi W, Mizuno K et al 1997 An expression profile of genes in human retina and isolation of a complementary DNA for a novel rod photoreceptor protein. Invest Ophthalmol Vis Sci 38:2576–2585

Sinha S, Sharma A, Agarwal N, Swaroop A, Yang-Feng TL 2000 Expression profile and chromosomal location of cDNA clones, identified from an enriched adult retina library. Invest Ophthalmol Vis Sci 41:24–28

Stohr H, Mah N, Schultz HL, Gehrig A, Frohlich S, Weber BH 2000 EST mining of the UniGene dataset to identify retina-specific genes. Cytogenet Cell Genet 91:267–277

Strausberg RL, Simpson AJ, Wooster R 2003 Sequence-based cancer genomics: progress, lessons and opportunities. Nat Rev Genet 4:409–418

Swaroop A, Zack DJ 2002 Transcriptome analysis of the retina. Genome Biol 3: REVIEWS1022.

Velculescu VE, Zhang L, Vogelstein B, Kinzler KW 1995 Serial analysis of gene expression. Science 270:484–487

Wang WH, McNatt LG, Shepherd AR et al 2001 Optimal procedure for extracting RNA from human ocular tissues and expression profiling of the congenital glaucoma gene FOXC1 using quantitative RT-PCR. Mol Vis 7:89–94

Whitney AR, Diehn M, Popper SJ et al 2003 Individuality and variation in gene expression patterns in human blood. Proc Natl Acad Sci USA 100:1896–1901

Wilson AS, Hobbs BG, Shen WY et al 2003 Argon laser photocoagulation-induced modification of gene expression in the retina. Invest Ophthalmol Vis Sci 44:1426–1434

Wirtz MK, Xu H, Rust K, Alexander JP, Acott TS 1998 Insulin-like growth factor binding protein-5 expression by human trabecular meshwork. Invest Ophthalmol Vis Sci 39:45–53

Wistow G, Bernstein SL, Wyatt MK et al 2002 Expressed sequence tag analysis of human retina for the NEIBank Project: retbindin, an abundant, novel retinal cDNA and alternative splicing of other retina-preferred gene transcripts. Mol Vis 8:196–204

Yoshida S, Yashar BM, Hiriyanna S, Swaroop A 2002 Microarray analysis of gene expression in the aging human retina. Invest Ophthalmol Vis Sci 43:2554–2560

Yoshii T, Tamura K, Ishiyama I 1992 Presence of a PCR-inhibitor in hairs (Japanese). Nippon Hoigaku Zasshi 46:313–316

Yoshii T, Tamura K, Taniguchi T, Akiyama K, Ishiyama I 1993 Water-soluble eumelanin as a PCR-inhibitor and a simple method for its removal (Japanese). Nippon Hoigaku Zasshi 47:323–329

Yoshimura N, Kikuchi T, Kuroiwa S, Gaun S 2003 Differential temporal and spatial expression of immediate early genes in retinal neurons after ischemia–reperfusion injury. Invest Ophthalmol Vis Sci 44:2211–2220

Yu J, Othman MI, Farjo R et al 2002 Evaluation and optimization of procedures for target labeling and hybridization of cDNA microarrays. Mol Vis 8:130–137

DISCUSSION

Swaroop: Initially, when you and I began studying the regulation of rhodopsin we identified NRL, then CRX. Since then, many transcription factors and their interacting proteins (e.g. RX, QRX, BAF and FIZ) have been implicated. I'm sure there are many more factors and interacting proteins that will soon be found. How do we go about looking for the *in vivo* relevance of all of these interactions?

Zack: My dream is to be able to take a non-photoreceptor cell such as a 293 cell, and transfect a subset of factors so that we could turn on the endogenous rhodopsin gene that is not normally expressed. It is not clear whether this is theoretically possible. This is one approach: to find the subset of proteins that will allow that. Another avenue that might prove fruitful is chip–chip assays, which involves bringing down the chromosomes with an antibody against a specific transcription factor, and then you figure out where they are binding the chromosome.

Swaroop: Unless there is a master gene that turns all the genes on, it will be hard to take 15 different genes and put them in a cell!

Zack: We could express perhaps half a dozen genes in 293 cells and get expression.

Thompson: Some of them may already be there.

Zack: The more interesting ones won't be.

Daiger: One of the interests in genetics is applying computational methods to detect binding sites for expression factors. This runs into the problem that most of the sites are 6 or 8 nucleotides, hence they are found throughout the genome. But if you look at these as a cluster of required sites for this whole set of complementary expression factors, can you have more success at a computational level in finding the regulatory regions around rhodopsin or the other opsins?

Zack: I hope so. When we initially compared rhodopsins from mouse, human and bovine, we were fortunate that the RER locus control region came out of a sequence comparison. We have tried doing this a lot since then, but it hasn't worked at all. Microarrays are likely to be an effective way to address this because we can use them to identify over 200 expressed genes in the retina. After we get these putative sites the question then remains about how we test them and show that they are important *in vivo*.

Farber: One of ways to address the issue Anand Swaroop raised is *ex vivo* transfection of *Xenopus* embryos to see whether you get the same results as transfecting cell lines. We have been doing this in collaboration with Barry Knox. Once we have *ex vivo* transfections we can do transgenics and see what happens there.

Zack: Any of these systems deals with basically one gene at a time, but we want to be able to address these issues at a multifactorial, complex level. We want to integrate the modifier genes so we can understand the complex matrix of interactions.

Swaroop: The issue is one of evaluating the critical *trans*-factors for regulation *in vivo*, out of the many that we know of. One way to do this might be *in vivo* footprinting.

Zack: That is a different method, but it is basically looking at the same thing as chromatin immunoprecipitation.

Daiger: What are the sizes of the families in which you found the pathogenic variants?

Zack: Unfortunately they are very small — they are singletons. Of the three, one person had a brother in India who could not be found. It has been very hard.

Thompson: I would like to ask about your microarray work. You have 10 000 genes on your array. What is the redundancy? Are there 10 000 unique sequences on your slide? How different are your slides from the cDNAs that other people have analysed, and does this relate to why different groups have seen different things with array methods?

Zack: There is about 5% redundancy. One approach we have that is different from what Anand Swaroop and you have done is that all the spots on your slides have been sequenced in house whereas we got sequences from many different places. We didn't have the resources to sequence all 10 000 again. Our approach is to identify the interesting differentially expressed sequences, and then we go back to that spot and sequence it. We think this is a good way of correcting errors in the original sequences. In terms of why different groups see different things, we have mentioned the differences between the genes on the arrays. I am sure there is significant overlap as well as significant non-overlap.

From disease genes to cellular pathways: a progress report

J. Yu*†, A. J. Mears*, S. Yoshida*, R. Farjo*, T. A. Carter‡, D. Ghosh§, A. Hero†¶, C. Barlow‡, A. Swaroop*‖[1]

*Departments of *Ophthalmology and Visual Sciences, †Biomedical Engineering, §Biostatistics, ¶Statistics and ‖Human Genetics, University of Michigan, Ann Arbor, MI 48105-0714 and ‡The Salk Institute for Biological Studies, Laboratory of Genetics, 10010 North Torrey Pines Road, La Jolla, CA 92037, USA*

Abstract. Mutations in a large number of retinal and retinal pigment epithelium (RPE) expressed genes can lead to the degeneration of photoreceptors and consequently the loss of vision. The genetic and phenotypic heterogeneity of retinal dystrophies poses a complex problem with respect to rational development of therapeutic strategies. Delineation of physiological functions of disease genes and identification of pathways that lead to disease pathogenesis represent essential goals towards developing a systematic and global approach to gene-based treatments. We are interested in identifying cellular pathways that are involved in photoreceptor differentiation, function and degeneration. We are, therefore, generating comprehensive gene expression profiles of retina and RPE of humans and mice using both cDNA- and oligonucleotide-based (Affymetrix) microarrays. Because of the under-representation of retinal/ RPE genes in the public databases, we have constructed several unamplified cDNA libraries and produced almost twenty thousand expressed sequence tags (ESTs) that are being printed onto glass slides ('*I-Gene*' microarrays). In this presentation, we will report the microarray analysis of the rodless (and cone-enhanced) retina from the *Nrl*-knockout mouse as a paradigm to initiate the identification of cellular pathways involved in photoreceptor differentiation and function.

2004 Retinal dystrophies: functional genomics to gene therapy. Wiley, Chichester (Novartis Foundation Symposium 255) p 147–164

Background and basic concepts

Retinal dystrophies (RD) comprise a group of clinically and genetically heterogeneous retinal disorders, which typically result in the degeneration of

---

[1]This paper was presented at the symposium by A. Swaroop to whom all correspondence should be addressed.

photoreceptors followed by the impairment or loss of vision. To date, the online retinal information network (RetNet, *http://www.sph.uth.tmc.edu/Retnet*) has listed over 130 loci associated with retinal dystrophies. RD is a major cause of blindness in the industrialized world and is, for the most part, currently untreatable. Retinitis pigmentosa (RP) primarily causes rod photoreceptor degeneration and early symptoms include night blindness and loss of peripheral vision. The prevalence of RP is approximately 1/3000, with a total of over 1.5 million people affected worldwide (Saleem & Walter 2002). In contrast, cone dysfunction occurs early during the progression of cone or cone-rod dystrophies (CRD), thereby affecting visual acuity and colour vision. Leber congenital amaurosis (LCA) is the most common cause of congenital visual impairment with age of onset in infants or children. LCA accounts for 5–10% of all retinal dystrophies and is perhaps the most severe RD. Age-related macular degeneration (AMD) is highly prevalent in the elderly population, accounting for 22% of monocular blindness and 75% of legal blindness in adults over age 50 in the USA (Klein et al 1995). It preferentially affects the macular region, leading to loss of central vision and visual acuity. Unlike other forms of RD, AMD is the culmination of a complex interplay of genetic and non-genetic components. The complexity afforded by the considerable genetic heterogeneity in RD has greatly hindered the application of gene-based therapies; nonetheless, all of these diseases result in the same fate, i.e. the death of the photoreceptors.

A number of innovative strategies have been employed with the objectives of slowing down, preventing, or even reversing photoreceptor cell death in RD. One approach of circumventing the heterogeneity of RD is symptom-based disease treatments without correcting underlying genetic defects. To restore sight in highly visually handicapped individuals, several research groups are working on the development of electronic photoreceptor prosthesis (Zrenner et al 2001, Hammerle et al 2002) and cell/tissue transplantations (Otani et al 2002, Radner et al 2002, Semkova et al 2002). However, these strategies are currently limited due to issues regarding biocompatibility, stability and longevity of transplants. Another generic approach involves the use of growth or survival factors (LaVail et al 1998). In any event, the need for understanding both the physiological function of disease genes and the cellular processes leading to photoreceptor degeneration is inescapable.

Gene-based therapy seeks to rescue retinal diseases by correcting the underlying genetic defect or a consequent physiological deficiency. Over 80 genes have been associated with retinal dystrophies (Bessant et al 2001, Saleem & Walter 2002), including the neural retinal leucine zipper (*NRL*) gene, *NR2E3* (nuclear receptor subfamily 2, group E, member 3), *PDE6B* (phosphodiesterase 6B, cGMP-specific, rod, beta), *CRX* (cone-rod homeobox) and *RHO* (rhodopsin). To rescue RD, numerous researchers have attempted to deliver a functional copy

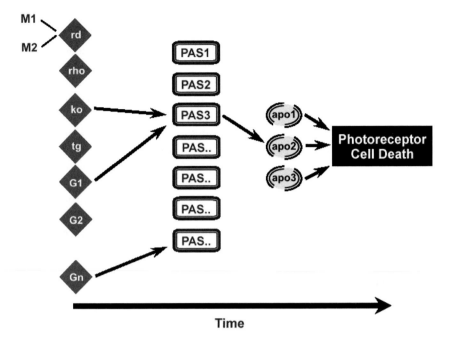

FIG. 1.   The progression of disease in retinal dystrophies: from genes to pathways. This schematic representation shows our approach for gene-based therapy that focuses on the convergence of different pre-apoptotic cellular pathways in time, in order to develop novel therapeutic targets for several forms of RD. In a majority of retinal dystrophies, the photoreceptors die by apoptosis. Mutations in hundreds of genes may disrupt the cellular homeostasis and selected signalling pathways. M1, M2 represent different mutations in the same gene (*rd*), and the blue squares indicate various 'disease' genes. The response of photoreceptors to the presence of a mutation is predicted to converge on a few pre-apoptotic signalling pathways (PAS1,2,3 . . . indicates pre-apoptic signals) that lead eventually to photoreceptor cell death via apoptotic pathways (apo1–3). In this model, various pre-apoptotic signals (PAS) would be ideal targets for drug discovery.

of the mutant gene into photoreceptor cells using viral-based vectors (Bennett et al 1998, Cheng et al 2002). However, gene transfer technology faces a number of hurdles, including the sheer number of distinct targets that need to be addressed due to the heterogeneity of RD, and issues regarding the safety and efficacy of such vectors.

An alternative approach that we advocate is a therapeutic design based on the understanding of the cellular pathways leading to photoreceptor cell death (Fig. 1). Although a large number of retinal and retinal pigment epithelium (RPE) expressed genes can lead to RD, studies have shown that only a few common cellular pathways are involved in disease progression and the photoreceptor cells in many, if not all, forms of RD die via apoptosis (Travis 1998). Pharmacological approaches have been advanced to slow photoreceptor degeneration through the introduction of growth and survival factors (LaVail et al 1998, Liang et al 2001, Tao et al 2002). Unfortunately, most experiments were only able to slow cell death for a week to a month, possibly due to the irreversible stage of disease by the time apoptotic pathways are induced. In order to devise a therapeutic strategy that targets multiple forms of RD prior to the induction of massive photoreceptor cell death, we are elucidating the common pathways of photoreceptor degeneration at a pre-apoptotic stage of disease. As illustrated in Fig. 1, pathways of disease pathogenesis initiated by different mutant gene products (or the lack thereof) must converge over time and follow limited routes to cell death. Therefore, temporal profiling of gene expression in normal developing, mature and ageing retinas and in retinal degeneration mouse models should lead to the identification of common pre-apoptotic signals (PAS) that can be targeted for drug discovery. A crucial aspect of this approach is the understanding of normal differentiation and function of rods and cones since it serves as the baseline against which abnormal changes may be recognized.

We propose that the adaptive response of the retinal neurons or RPE to disease or ageing is reflected by modulation of specific cellular pathways and, consequently, changes in gene expression. Profiling of diseased or ageing retina or RPE from humans and mice will facilitate the identification of these pathways. In this manuscript, we will primarily focus on the regulatory networks of photoreceptor development and function in the context of the transcription factor *Nrl*, using the *Nrl*$^{-/-}$ mouse as a paradigm.

Nrl: an essential transcription factor for rod development and function

The *Nrl* gene, encoding a basic motif leucine zipper protein of Maf-subfamily, was initially identified from a subtracted retinal library (Swaroop et al 1992). It showed a highly restricted pattern of expression, primarily in rod photoreceptors (Farjo et al 1993, Swain et al 2001). Six phosphorylated isoforms of *Nrl* have been identified in rod but not cone photoreceptor nuclei (Swain et al 2001). The *Nrl* protein can positively regulate rhodopsin gene expression by binding to an extended AP-1-like sequence element (called NRE) in the upstream promoter region (Kumar et al 1996, Rehemtulla et al 1996). Further studies indicated that *Nrl* regulates several other rod genes, and can interact with other transcriptional factors, such as *Crx*, in the regulation of retinal expressed genes (Chen et al 1997, Mitton et al 2000, Lerner et al 2001). Mutations in the human *NRL* gene have been associated with autosomal dominant RP (Bessant et al 1999, 2000, Martinez-Gimeno et al 2001, DeAngelis et al 2002). Interestingly, 5 of the 6 currently identified mutations alter the residues S50 and P51, resulting in possibly hypermorphic alleles of NRL and suggesting their functional importance.

To define the role of *Nrl* in photoreceptor development and function, the *Nrl* gene was deleted in mice by homologous recombination (Mears et al 2001). Since *Nrl* plays a key role in the regulation of rod-specific genes, it was anticipated that the deletion of *Nrl* would affect rod photoreceptors. Surprisingly, the *Nrl*$^{-/-}$ mouse retina is functionally rodless. The knockout retina has abnormal histology, with rosettes and whorls within the outer nuclear layer. Only 20% of photoreceptors elaborate outer segments, most of which have abnormal disk morphology. Electroretinogram (ERG) recording revealed no scotopic response and detected a light-adapted b-wave of two to three times larger amplitude in knockout than that of wild-type retina, demonstrating the absence of rod function and an enhanced cone function. Using monochromatic stimuli of 400 nm or 530 nm, this large b-wave amplitude is explained by increased S-cone activity. Preliminary gene expression analysis revealed an absence of rod-specific transcripts, and an increase in the expression of cone-specific genes (Mears et al 2001). Dramatic retinal changes observed in this mouse establish it as an excellent model for expression profiling corresponding to different pathways associated with rod and cone development and function. We propose genes with reduced expression in the *Nrl*$^{-/-}$ retina relative to normal would be associated with rod-signalling pathways, while those with augmented expression relate to cone function.

Microarray analysis

High-throughput technologies, including cDNA microarrays and Affymetrix GeneChips have made large-scale gene expression studies of retinal tissues readily achievable (Farjo et al 2002, Yoshida et al 2002, Swaroop & Zack 2002). Microarrays allow us to investigate changes in expression at a genome scale in a single experiment. This approach is limited only by the number and types of genes represented on the arrays. In addition to being a powerful gene-discovery tool in the identification of candidate genes, microarrays may shed considerable light on the cellular pathways of the tissue under study (Livesey et al 2000, 2002). A schema of microarray analysis is presented in Fig. 2. Although Affymetrix technology is relatively well developed, with appropriate quality controls, standard data pre-processing and ready-to-use data analysis software, its application to our studies is limited by the under-representation of retinal expressed genes on their GeneChips. For comprehensive profiling, customized *I-Gene* cDNA microarrays were also utilized. These arrays were generated by printing retina/eye-expressed genes and expressed sequence tags (ESTs) obtained from a variety of cDNA libraries (*www.umich.edu/ ~ igene/*; Yu et al 2003) onto glass

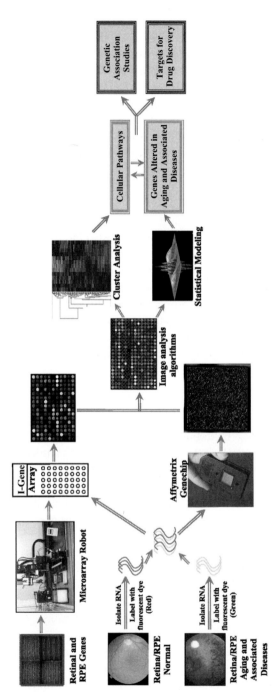

FIG. 2. Comprehensive gene profiling of control and mutant retinas using Affymetrix GeneChips and custom *I-Gene* microarrays. Temporal expression profiling followed by statistical modelling and cluster analysis can lead to the identification of pathways and molecular targets.

slides using a robotic micro-arrayer (Farjo et al 2002, Yu et al 2002). For these high-throughput studies, total RNA was isolated from either control (normal) or experimental (diseased or ageing) retinas, labelled with fluorescent dyes and hybridized to either Affymetrix GeneChips or *I-Gene* microarrays (Fig. 2). Image analysis and statistical modelling were employed to identify differentially expressed genes between control and experimental samples. Clustering algorithms were used to group co-expressed genes under different experimental conditions, which might lead to the identification of functional/regulatory networks and pathways (Fig. 2). We have used gene profiling of retinas from the normal and *Nrl*-knockout mice as a paradigm and to establish the proof of principle.

Affymetrix GeneChip study

Gene profiling of postnatal day 2 (PN2), PN10 and 2 month-old retinas from the control and *Nrl*-knockout mice using mouse GeneChips showed approximately equal number of up- or down-regulated genes at each time point (data not shown). At PN2, only 6 genes are found to be differentially expressed, compared with 74 at PN10 and 136 at 2 months. As predicted, several rod photoreceptor-specific genes, including rhodopsin (*Rho*) and rod transducin alpha (*Gnat1*), were found to be greatly under-expressed in the knockout mouse, while cone genes, such as S-opsin (*Opn1sw*) and cone transducin alpha (*Gnat2*), are up-regulated. Quantitative real-time PCR (qRT-PCR) analyses of almost 50 genes have validated gene expression changes revealed by GeneChips; qRT-PCR profiles of four genes are shown in Fig. 3. More than 20% of differentially expressed transcripts were unknown ESTs. These are of considerable interest, as they may represent novel retinal dystrophy candidate genes or lead to the elucidation of specific cellular pathways associated with photoreceptor differentiation and function. Clusters of differentially expressed genes may also provide insights into pathways and functional networks (Fig. 4).

I-Gene micoarray study

Gene expression of wild type and $Nrl^{-/-}$ mice retinas were compared at five developmental time points: PN0, PN2, PN6, PN10 and PN21. Custom *I-Gene* microarrays containing over 6500 eye/retina expressed genes and ESTs printed in duplicate were generated for hybridization (Figs 5A,B). Five replicates were performed for each stage utilizing labelled targets from different mice to reduce individual variance. Density plots of the log-ratios of gene expression in PN21 $Nrl^{+/+}$ and $Nrl^{-/-}$ mice retinas detected by five independent replicated experiments showed similar patterns of distribution. Log-ratios of all replicates are centred at 0, with most genes lying within −1 and +1 (Fig. 5C), suggesting

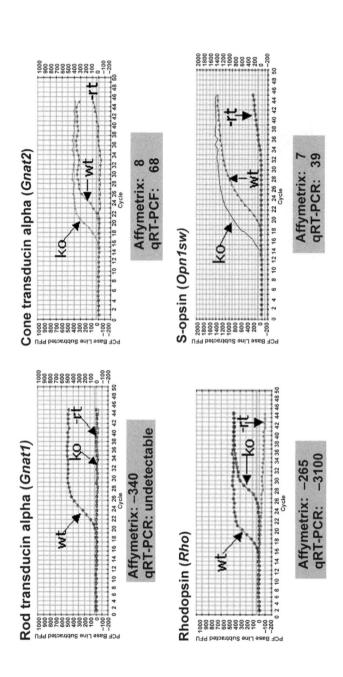

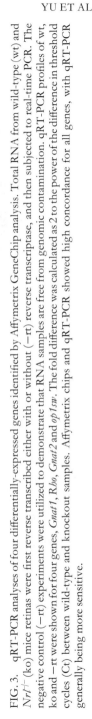

FIG. 3.   qRT-PCR analyses of four differentially-expressed genes identified by Affymetrix GeneChip analysis. Total RNA from wild-type (wt) and Nrt[−/−] (ko) mice retinas were first reverse transcribed either with or without (−rt) reverse transcriptase, and then subjected to real-time PCR. The negative control (−rt) experiments were utilized to demonstrate that RNA samples are free from genomic contamination. qRT-PCR profiles of wt, ko and −rt were shown for four genes, Gnat1, Rho, Gnat2 and op1sw. The fold difference was calculated as 2 to the power of the difference in threshold cycles (Ct) between wild-type and knockout samples. Affymetrix chips and qRT-PCR showed high concordance for all genes, with qRT-PCR generally being more sensitive.

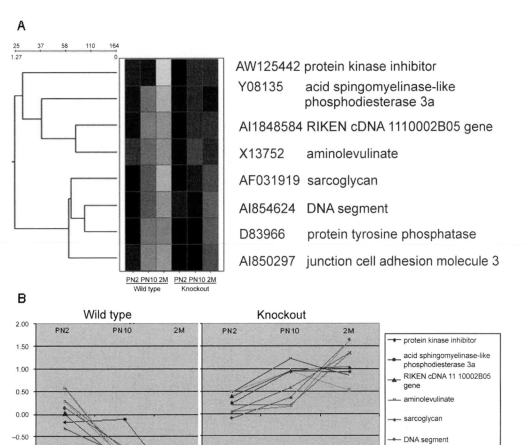

FIG. 4. Cluster analysis of the temporal expression profiles generated from Affymetrix GeneChips. (A) Representation of clustering analysis of differentially expressed genes. The data matrix was first standardized to z-score and hierarchical clustering analysis performed using the 'Euclidean distance' method. Colour-coding indicates relative expression: green being low, red high (this appears as grey scale on this black and white reproduction). Eight genes shown are clustered based on their similarity of expression profile, which is also graphically represented in (B), where z-scores (Y-axis) are plotted against time-point (X-axis).

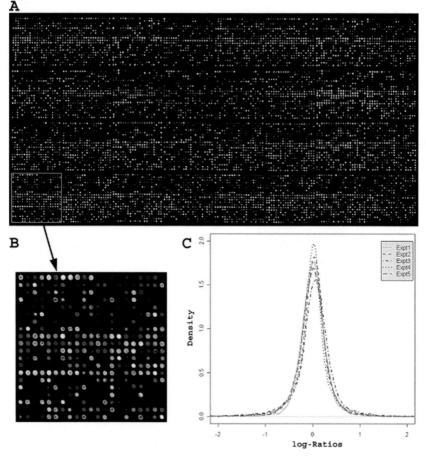

FIG. 5. *I-Gene* microarray and density plots of log-ratios in five replicate experiments. (A) A TIFF image of the Cy3 channel of an *I-Gene* microarrays containing over 6500 genes or ESTs printed in duplicate. False colour has been applied to indicate the intensity of hybridization, with black having no signal, blue low, red high, and white saturated (these appear as grey scales on this black and white reproduction). (B) Enlargement of the left lower corner grid of the array, showing uniform spot diameter, clear hybridization and low background signal. (C) Ratios of gene expression indicate the abundance of each gene in $Nrl^{-/-}$ mice retinas relative to $Nrl^{+/+}$ retinas. Smooth density plots of log-ratios shows that, in all replicates (expt1–expt5), log-ratios are centred at 0, with majority of spots lying between −1 and +1.

that the expression of a majority of genes is unaltered or minimally altered between the control and *Nrl*-knockout retinas. Microarrays tend to underestimate the true biological change and perhaps a log ratio threshold of less than 1 needs to be established. Statistical analysis of PN21 expression data identified 52 cDNAs, representing 39 unique genes, with the highest possibility of differential

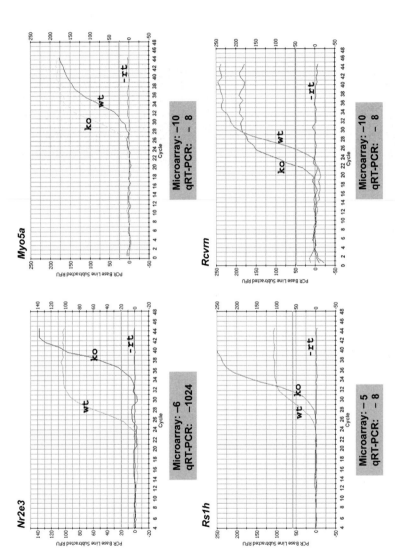

FIG. 6.   qRT-PCR validation of *I-Gene* microarray results: analysis of *Nr2e3*, *Rs1h*, *Myo5a*, and *Rcvrn* expression in retinas of wild-type (wt) and *Nrl*-knockout (ko) mice. Total RNA from wild-type and knockout mice retinas were first reverse transcribed either with or without (−rt) reverse-transcriptase, and then subjected to real-time PCR. The negative control (−rt) experiments were utilized to demonstrate that RNA samples are free from genomic DNA contamination. qRT-PCR tends to be more sensitive than the hybridization-based microarray experiments.

expression. Over 30% of these genes are known to play important roles in the retina; these include *Rho*, *Opn1sw*, *Gnat1*, *Gnat2*, *Nr2e3*, Retinoschisis 1 homolog (*Rs1h*), myosin 5a (*Myo5a*) and Recoverin (*Rcvrn*). qRT-PCR analyses validated these expression alterations (Fig. 6). Further examination of these differentially expressed genes suggests a bias in the utilization of the bone morphogenetic protein (Bmp) signalling pathway, Wnt/Ca^{2+} signalling pathway and the retinoic acid pathway between rods and cones (J. Yu, A.J. Mears and A. Swaroop, unpublished data).

Pathway consolidation

Affymetrix GeneChip studies, presented here, showed differential gene expression from PN2, PN10 to 2-month-old retinas, whereas *I-Gene* cDNA microarray data indicated alterations of signalling pathways in the PN21 knockout mice retinas. Systematic examination of gene expression levels at PN0, PN2, PN6, PN10 and PN21 followed by statistical analysis should further assist in the identification of genes that are downstream of *Nrl* in regulatory hierarchy and play key roles in photoreceptor differentiation and/or function. Clustering based on temporal expression profiles may identify coordinately regulated genes involved in rod and cone photoreceptor development. Since hypermorphic alleles of *Nrl* are predicted to cause retinal degeneration, the signalling pathways downstream of *Nrl* may also be studied in the context of other retinal degenerative mouse models.

Conclusions

Delineation of cellular pathways involved in photoreceptor differentiation and disease pathogenesis presents an attractive approach to identify targets for treatment of RD. In this presentation, we have used a single paradigm to illustrate our research approach and the focus on cellular pathways downstream of an important retinal gene. *Nrl* is a rod-specific transcription factor that is required for rod differentiation and regulation of rod-specific gene expression. Mutations in the human *NRL* gene have been identified in patients with autosomal dominant RP. The $Nrl^{-/-}$ mouse retina is rodless, with an increased number of functional S-cones. Using Affymetrix GeneChips and custom *I-Gene* cDNA microarrays, we have so far identified over 150 genes that are differentially expressed in the *Nrl*-knockout mouse retina as compared to controls. Several of these cDNAs represent novel genes that are attractive candidates for RD. Further characterization of differentially-expressed cDNAs should reveal direct or indirect targets of *Nrl* and assist in developing transcriptional regulatory hierarchy downstream of *Nrl*. Initial studies also suggest differential utilization of signalling pathways in rods and cones. Our investigations provide an initial framework for establishing pathway-based treatment strategies for retinal and macular diseases.

Acknowledgements

We thank Mohammad Othman and Sean MacNee for their advice and assistance. The research in our laboratory is supported by grants from the National Institutes of Health (EY11115 including administrative supplements, EY07961, and EY07003), The Foundation Fighting Blindness (Owings Mills, MD), Macula Vision Research Foundation (West Conshohocken, PA), Research to Prevent Blindness (New York, NY), Elmer and Sylvia Sramek Charitable Foundation (Chicago, IL), and Juvenile Diabetes Research Foundation (New York, NY).

References

Bennett J, Zeng Y, Bajwa R, Klatt L, Li Y, Maguire AM 1998 Adenovirus-mediated delivery of rhodopsin-promoted bcl-2 results in a delay in photoreceptor cell death in the rd/rd mouse. Gene Ther 5:1156–1164

Bessant DA, Payne AM, Mitton KP, et al 1999 A mutation in NRL is associated with autosomal dominant retinitis pigmentosa. Nat Genet 21:355–356

Bessant DA, Payne AM, Plant C, Bird AC, Swaroop A, Bhattacharya SS 2000 NRL S50T mutation and the importance of 'founder effects' in inherited retinal dystrophies. Eur J Hum Genet 8:783–787

Bessant DA, Ali RR, Bhattacharya SS 2001 Molecular genetics and prospects for therapy of the inherited retinal dystrophies. Curr Opin Genet Dev 11:307–316

Chen S, Wang QL, Nie Z et al 1997 Crx, a novel Otxl-like paired-homeodomain protein, binds to and transactivates photoreceptor cell-specific genes. Neuron 19:1017–1030

Cheng L, Chaidhawangul S, Wong-Staal F et al 2002 Human immunodeficiency virus type 2 (HIV-2) vector-mediated in vivo gene transfer into adult rabbit retina. Curr Eye Res 24:196–201

DeAngelis MM, Grimsby JL, Sandberg MA, Berson EL, Dryja TP 2002 Novel mutations in the NRL gene and associated clinical findings in patients with dominant retinitis pigmentosa. Arch Ophthalmol 120:369–375

Farjo Q, Jackson AU, Xu J et al 1993 Molecular characterization of the murine neural retina leucine zipper gene, Nrl. Genomics 18:216–222

Farjo R, Yu J, Othman MI et al 2002 Mouse eye gene microarrays for investigating ocular development and disease. Vision Res 42:463–470

Hammerle H, Kobuch K, Kohler K, Nisch W, Sachs H, Stelzle M 2002 Biostability of micro-photodiode arrays for subretinal implantation. Biomaterials 23:797–804

Klein R, Wang Q, Klein BE, Moss SE, Meuer SM 1995 The relationship of age-related maculopathy, cataract, and glaucoma to visual acuity. Invest Ophthalmol Vis Sci 36:182–191

Kumar R, Chen S, Scheurer D et al 1996 The bZIP transcription factor Nrl stimulates rhodopsin promoter activity in primary retinal cell cultures. J Biol Chem 271:29612–29618

LaVail MM, Yasumura D, Matthes MT et al 1998 Protection of mouse photoreceptors by survival factors in retinal degenerations. Invest Ophthalmol Vis Sci 39:592–602

Lerner LE, Gribanova YE, Ji M, Knox BE, Farber DB 2001 Nrl and Sp nuclear proteins mediate transcription of rod-specific cGMP-phosphodiesterase beta-subunit gene: involvement of multiple response elements. J Biol Chem 276:34999–35007

Liang FQ, Dejneka NS, Cohen DR et al 2001 AAV-mediated delivery of ciliary neurotrophic factor prolongs photoreceptor survival in the rhodopsin knockout mouse. Mol Ther 3: 241–248

160 DISCUSSION

Livesey FJ, Furukawa T, Steffen MA, Church GM, Cepko CL 2000 Microarray analysis of the transcriptional network controlled by the photoreceptor homeobox gene Crx. Curr Biol 10:301–310
Livesey R 2002 Have microarrays failed to deliver for developmental biology? Genome Biol 3:COMMENT2009
Martinez-Gimeno M, Maseras M, Baiget M et al 2001 Mutations P51U and G122E in retinal transcription factor NRL associated with autosomal dominant and sporadic retinitis pigmentosa. Hum Mutat 17:520
Mears AJ, Kondo M, Swain PK et al 2001 Nrl is required for rod photoreceptor development. Nat Genet 29:447–452
Mitton KP, Swain PK, Chen S, Xu S, Zack DJ, Swaroop A 2000 The leucine zipper of NRL interacts with the CRX homeodomain. A possible mechanism of transcriptional synergy in rhodopsin regulation. J Biol Chem 275:29794–29799
Otani A, Kinder K, Ewalt K, Otero FJ, Schimmel P, Friedlander M 2002 Bone marrow-derived stem cells target retinal astrocytes and can promote or inhibit retinal angiogenesis. Nat Med 8:1004–1010
Radner W, Sadda SR, Humayun MS, Suzuki S, de Juan E Jr 2002 Increased spontaneous retinal ganglion cell activity in rd mice after neural retinal transplantation. Invest Ophthalmol Vis Sci 43:3053–3058
Rehemtulla A, Warwar R, Kumar R, Ji X, Zack DJ, Swaroop A 1996 The basic motif–leucine zipper transcription factor Nrl can positively regulate rhodopsin gene expression. Proc Natl Acad Sci USA 93:191–195
Saleem RA, Walter MA 2002 The complexities of ocular genetics. Clin Genet 61:79–88
Semkova I, Kreppel F, Welsandt G et al 2002 Autologous transplantation of genetically modified iris pigment epithelial cells: a promising concept for the treatment of age-related macular degeneration and other disorders of the eye. Proc Natl Acad Sci USA 99:13090–13095
Swain PK, Hicks D, Mears AJ et al 2001 Multiple phosphorylated isoforms of NRL are expressed in rod photoreceptors. J Biol Chem 276:36824–36830
Swaroop A, Zack DJ 2002 Transcriptome analysis of the retina. Genome Biol 3:REVIEWS1022
Swaroop A, Xu JZ, Pawar H, Jackson A, Skolnick C, Agarwal N 1992 A conserved retina-specific gene encodes a basic motif/leucine zipper domain. Proc Natl Acad Sci USA 89:266–270
Tao W, Wen R, Goddard MB et al 2002 Encapsulated cell-based delivery of CNTF reduces photoreceptor degeneration in animal models of retinitis pigmentosa. Invest Ophthalmol Vis Sci 43:3292–3298
Travis GH 1998 Mechanisms of cell death in the inherited retinal degenerations. Am J Hum Genet 62:503–508
Yoshida S, Yashar BM, Hiriyanna S, Swaroop A 2002 Microarray analysis of gene expression in the aging human retina. Invest Ophthalmol Vis Sci 43:2554–2560
Yu J, Othman MI, Farjo R et al 2002 Evaluation and optimization of procedures for target labeling and hybridization of cDNA microarrays. Mol Vis 8:130–137
Yu J, Farjo R, MacNee SP, Baehr W, Stambolian DE, Swaroop A 2003 Annotation and analysis of 10,000 expressed squence tags from developing mouse eye and adult retina. Genome Biol 4:R65 (open access full text at *http://www.genomebiology.com/2003/4/10/R65*)
Zrenner E, Gekeler F, Gabel VP et al 2001 Subretinal microphotodiode array as replacement for degenerated photoreceptors? (German) Ophthalmologe 98:357–363

DISCUSSION

McInnes: When is *Nrl* turned on developmentally?

Swaroop: By RT-PCR we can detect it around E16.5 in embryonic mouse retina, but by Northern analysis it is more like E18.5. The antibodies we have currently pick up another protein called p45, which is present in all developing neurons. We are currently generating additional, more specific antibodies.

McInnes: Is p45 a product of the *Nrl* gene?

Swaroop: No, it is encoded by a different gene expressed probably in all neural cells. It is antigenically similar.

Hauswirth: In the *Nrl* knockout mouse, is the enhanced photopic ERG amplitude due to the presence of more cones? Or is there a higher response from the cones that are there?

Swaroop: The ERG studies show a higher response but there is more S opsin. The outer segments of Cods (cone–rod hybrids) that are there have S opsin.

Hauswirth: So is the extra amplitude coming from 'Cods', not from a conversion of rods to real cones?

Swaroop: We don't know what these Cods are. We use this term because we don't want to call them cones. It is too early to say whether they are rods converted to cones. We are working on it.

Hauswirth: What about the rest of the phototransduction cycle in cones?

Swaroop: It is all present in these Cod outer segments. All rod-specific genes have been switched off. None of the rod-specific proteins are expressed, whereas every cone-specific phototransduction protein that we have looked at is expressed at high levels.

Hauswirth: So if you want to preserve cone function in humans you just need to knock out *NRL*!

Bok: Anand Swaroop, did you say that the photoreceptors in this knockout mouse do not die?

Swaroop: The function of these cones is bizarre, and we do not see any large-scale change in the thickness of the outer nuclear layer at least for 6 months. So there is minimal cell loss during this period.

Bok: I presume the reason that Ed Stone and others looked at S-cone enhanced syndrome was because there is some sort of disease process in those retinas. Is there a cell loss, or is it just bizarre physiology?

Swaroop: I think there may be cell death in the *rd7* mouse.

Farber: The *rd7* mouse has a mutation in the *PNR/Nr2e3* gene. *NR2E3* mutations in humans cause enhanced S-cone syndrome (ESCS).

Bok: Do those humans lose cells?

Farber: No.

Bird: They have a restricted form of retinitis pigmentosa, and different mutations in the same gene cause a very severe form of retinal dystrophy.

Swaroop: My understanding is that there is some degeneration of photo-receptors in ESCS.

Dryja: They are different mutations. These are knockout mice, and all the humans are dominant missense mutations.

Bok: So are you talking about a gain of function in humans?

Swaroop: No. The human mutations in *NR2E3* are also loss of function. Why do *NR2E3* mutations lead to retinal degeneration, whereas we don't see this in the *Nrl* knockout mice? We have only looked up to six months. The mice are now two years old and we are working with Dr Paul Sieving to examine the retina of older mice by histology and ERG to figure out whether there has been any loss of cones. We have done some histology in mice older than 6 months and the outer nuclear layer is thinner. I don't know whether there is slow loss of cone function, but we are working on it. I must state that these may not be real cones because they do express some rod markers. Our collaborator, Dr David Hicks, has two antibodies, Ret P3 and L1, which appear to specifically recognize rods, not cones. These two antibodies recognize antigens in the knockout retina. In addition, Dr Enrica Strettoi has observed that the synaptic connections these Cods are forming are also apparently different from the normal rod and cone connections. According to Dr Ed Pugh, the Cods function as cones.

Farber: Many years ago we worked a lot with ground squirrels, and found that they had some cells that were intermediate between rods and cones. They all happened to be S cone cells. It might be worth looking here.

Swaroop: Maybe they don't have NRL, and that is why they are all S cones.

Zack: Have you used arrays on the *Rd7* mouse to complement these?

Swaroop: Yes. We have done two time points but the data have not been analysed yet.

Bhattacharya: I have a general question about the microarray data. What is your feeling about the level of variability seen from one experiment to another?

Swaroop: The correlation coefficient we get with Affymetrix GeneChips is over 0.99. If the same person dissects the retina and at the same time of the day the variability is minimum. If you take another knockout mice you see a little more variability. In slide microarrays we get closer to 0.98, so there is a little bit more variability in these. Even in slide microarrays there are ways to normalize the data. We are working with data-driven normalization. Rather than a global normalization of signals over the whole slide we do this on the basis of the data on each slide. This helps a great deal.

Aguirre: In an earlier talk Donald Zack showed a variability that was mainly patient related rather than age related. What are the prospects for looking at microarray data on patients?

Swaroop: I was talking primarily about mouse, where the data are very clean. We have done human studies with eight Affymetrix chips for young and eight Affymetrix chips with old retina, and there is a huge amount of variability within the samples. You have to throw away many of the data that may be real but we can't be confident. I tend to be very conservative. This variability could be because of inherent variations in humans or because of many other factors, including tissue collection time and tissue preservation.

Cremers: There was a recent paper in *Science* showing the variance of different genes (Yan et al 2002). They showed that in families expression levels could vary in normal individuals two- to fourfold.

Swaroop: That is why we chose to work in mice.

Cremers: Why do you think it is different in mice?

Swaroop: Because we are working with isogenic strains and we are controlling the sample preparation more carefully. With mice the variation is very low if the same strains are used. In humans there are many confounding factors.

Thompson: Are you using gender matching in your analysis?

Swaroop: Yes. My feeling is that there will be some genes that will show large variation, but most of the genes do not change. Once we define the baseline expression profile of all genes it will be easier.

Bok: You would do a service to all of us to figure out what the gender differences are so we can subtract these out.

Swaroop: We are working on many aspects of microarray data with respect to retinal biology. We are also working on the development of databases and web-based sharing of information. We hope to make all of our data available on our website.

Bhattacharya: If we identify a mutation in a novel gene in humans, we may want to look at the disease biology (the impact of the mutation and how it might lead to cell death) through microarray techniques. If there is a huge amount of variation, would it be worthwhile generating a mouse model for each human gene and then studying the disease biology in the mouse?

Swaroop: From my own experience, the human work with microarrays has been very frustrating. Dr Shigeo Yoshida, a Japanese postdoc in my lab, had worked 16 h days for two years and produced few useful data, ultimately. For the postdoc's sake it might be better to use mice than humans, at least initially. I would also advise people not to do just one time point: I would look at the progression of disease in the mouse model and pick four or five time points. It is a lot of work, but the information gained is very valuable.

Thompson: In terms of the human data, once we can get a line on more pathways affected in retinal degeneration, so we are not looking at everything but just focusing on one pathway at a time, then it will be easier to see important changes.

Swaroop: Once you have defined the pathways then you can fit in the data you get from human studies. It is much easier.

Zack: In terms of variability in different genes, I agree. In our experiments it turns out that rhodopsin is one of the most variable genes in the retina. Because rhodopsin is expressed at such a high level it is very easy to measure accurately on an array, but the level can vary by sixfold just at the RNA level in age-matched individuals.

Bird: Does this vary by time of day?

Zack: We have too few data to answer that. But in mouse models rhodopsin is not one of the genes that is subject to significant circadian regulation.

Swaroop: That is a good point. With all our mice we dissect them between 12 and 2 pm because we are not sure whether this is a significant factor or not. I think it probably does matter. As long as everything is kept the same the variation is lower.

Reference

Yan H, Yuan W, Velculescu VE, Vogelstein B, Kinzler KW 2002 Allelic variation in human gene expression. Science 297:1143

Prospects for gene therapy

Robin R. Ali

Molecular Genetics, Institute of Ophthalmology, University College London, Bath St, London EC1V 9EL, UK

Abstract. Inherited retinal disease, which includes conditions such as retinitis pigmentosa (RP), affects about 1/3000 of the population in the Western world. It is characterized by gradual loss of vision and results from mutations in any one of 60 or so different genes. There are currently no effective treatments, but many of the genes have now been identified and their functions elucidated, providing a major impetus to develop gene-based treatments. Many of the disease genes are photoreceptor- or retinal pigment epithelium (RPE) cell specific. Since adeno-associated viral (AAV) vectors can be used for efficient gene transfer to these two cell types, we are developing AAV-mediated gene therapy approaches for inherited retinal degeneration using animal models that have defects in these cells. The retinal degeneration slow (rds or $Prph2^{Rd2/Rd2}$) mouse, a model of recessive RP, lacks a functional gene encoding peripherin 2, which is a photoreceptor-specific protein required for the formation of outer segment discs. We have previously demonstrated restoration of photoreceptor ultrastructure and function by AAV-mediated gene transfer of peripherin 2. We have now extended our assessment to central visual neuronal responses in order to show an improvement of central visual function. The Royal College of Surgeons (RCS) rat, provides another model of recessive RP. Here the defect is due to a defect in *Mertk*, a gene that is expressed in the RPE and encodes a receptor tyrosine kinase that is thought to be involved in the recognition and binding of outer segment debris. The gene defect results in the inability of the RPE to phagocytose the shed outer segments from photoreceptor cells. The resulting accumulation of debris between the RPE and the neuroretina leads to progressive loss of photoreceptor cells. AAV-mediated delivery of *Mertk* to the RPE results in reduction of debris indicating that the phagocytosing function of the RPE is restored and delays the degeneration of the photoreceptor cells 3–4 months. Our results, along with those of other groups support the use of AAV vectors for the treatment of inherited retinal degeneration.

2004 Retinal dystrophies: functional genomics to gene therapy. Wiley, Chichester (Novartis Foundation Symposium 255) p 165–176

Inherited retinal degenerations, which occur with a frequency of around 1/3000, are a major cause of inherited blindness in the Western world. They may result from a defect in any one of over 60 different genes, many of which are either photoreceptor- or retinal pigment epithelium (RPE)-specific (Bessant et al 2001). Although there are no effective treatments to date, a variety of novel therapeutic strategies are being investigated. These include administration of recombinant neurotrophic factors, photoreceptor, RPE or stem cell transplantation, and gene therapy. Of these different approaches, the most significant progress has probably been made with regard to gene therapy. Efficient *in vivo* gene transfer to photoreceptor cells and the RPE following subretinal injection of adeno-associated viral (AAV) vectors resulting in stable transgene expression has been demonstrated in a variety of animals, including rodents, dogs and primates with minimal inflammation and toxicity (Bennett et al 1999, Dudus et al 1999, Sarra et al 2002, Bainbridge et al 2003). The use of cell-specific promoters enables transgene expression to be restricted to either photoreceptors or RPE, providing an additional level of control. A number of recent studies have clearly demonstrated either retardation of photoreceptor cell loss or functional improvement in animal models of retinal degeneration following AAV-mediated gene transfer (McGee Sanftner et al 2001, Liang et al 2001, Acland et al 2001, Ali et al 2000). Here I will outline some of our work on the development of gene therapy for inherited retinal degeneration and compare the efficacy of treating of photoreceptor and RPE-specific defects.

We have focused most of our efforts on developing gene replacement therapy for inherited retinal degeneration using $Prph2^{Rd2/Rd2}$ mice. These animals, formerly known as *rds* (retinal degeneration slow) mice are a well-characterized model of retinal degeneration. They are homozygous for a null mutation in the $Prph2$ gene, encoding the structural glycoprotein, peripherin 2, which is essential for outer-segment formation (Travis et al 1991). Due to the failure to develop photoreceptor outer segments, $Prph2^{Rd2/Rd2}$ mice have extremely limited electroretinogram (ERG) responses. This facilitates reliable quantification of functional improvement following treatment making the $Prph2^{Rd2/Rd2}$ mouse a very useful tool for assessing the efficacy of gene therapy protocols. We have demonstrated improvement of photoreceptor ultrastructure at a number of time points following injection of an AAV vector carrying a peripherin 2 cDNA driven by a rhodopsin promoter (AAV.rho.rds) (Ali et al 2000, Sarra et al 2001). We have now analysed treated animals at weekly intervals using ERG (Schlichtenbrede et al 2003a,b). We have analysed ERG changes in a series of increasing stimulus intensities, using trace pattern and b-wave amplitude as indicators of functional improvement at several time points following treatment. Although there is some inter-animal and test–retest variability, the ERG has proven to be very reliable tool in assessing therapeutic benefit. In the absence of treatment, we found the concordance between the b-wave amplitudes of the two eyes of a single animal was very high, both in normal and $Prph2^{Rd2/Rd2}$ mice. Another phenomenon demonstrated in the recordings was the variability in rates of degeneration in different mice. This was in contrast to non-degenerating mice (wild-type CBA), where considerably less variation was seen over large a number of animals. The difference in variability may be explained by a difference in genetic heterogeneity. The $Prph2^{Rd2/Rd2}$ mice are on a CBA background but are not congenic. This is consistent with the observation in humans that within families manifesting the same degeneration and the same gene defect there can be a very large variation in the rate of degeneration and vision loss (Apfelstedt-Sylla et al 1995).

The b-wave pattern in the animals treated with AAV.rho.rds resemble that of a normal animal. Furthermore the presence of oscillatory potentials on rise of the b-wave indicate that the transverse connections of the outer retina involving amacrine cells are also functioning. The average b-wave amplitude in the rescued retina following single time point injection at 4 weeks post injection is 97 mV at the 100 mcds/m^2 stimulus. For a wild-type mouse (CBA) the b-wave amplitude at the same intensity is around 400 mV. Thus the b-wave amplitude in treated animals is approximately 25% of that in normal wild-type mice. Given the subretinal injection technique reaches approximately 50% of retina and about 70% of photoreceptors are transduced (Sarra et al 2002), the degree of the physiological rescue is consistent with the anticipated number of transduced photoreceptors.

We have established a positive correlation between the area of retina transduced and the magnitude of the ERG. This is reflected in the difference between the b-wave amplitudes following single time point and repeated injections. The average b-wave amplitude was 90 mV and 140 mV for the single time point and repeatedly injected animals respectively. This corresponds to a functional rescue over the untreated fellow eye of 80% and 250% respectively.

We have also evaluated the impact of gene replacement therapy on higher visual function. Demonstrating an improvement of central visual responses and correlating this to changes in the retina is a critical step for validating gene therapy approaches for inherited retinal disease. The effect of treatment on higher visual function was assessed by recording from central visually responsive neurons in the superior colliculus and improvements were correlated in individual animals with retinal function and histological and biochemical changes. Although gene replacement therapy only partially restores photoreceptor morphology, it results in a 300% increase of the visual cycle protein rhodopsin, leading to retinal function improvement, reflected by a 250% higher b-wave amplitude. This corresponds to 15% of the rhodopsin levels in normal animals and 25% of the normal b-wave amplitude. $Prph2^{Rd2/Rd2}$ mice with improved ERGs also had significantly higher central visual responses (166% increase at 24 cd/m^2). These findings suggest that gene replacement therapy leading to even relatively modest structural improvement may result in improved visual function.

Despite the clear functional improvement, beneficial effects of treatment appear to be transient. The onset of the rescue is consistent with the time course of AAV expression with a lag of approximately two weeks before effective transgene expression levels can be recorded. The improvement in ERG is clear and consistent between 3–12 weeks for all animals with repeated injections. The longest period of significant functional benefit is around 14 weeks. At later time points the differences between treated and untreated eyes are no longer significant. We have also observed a decline in ultrastructural quality of the outer segments over time even when we treat young animals. This might be explained either by increased disturbance of photoreceptor cell physiology over time, irrespective of outer segment induction, or by inappropriate transgene expression levels. We have also observed that the number of induced outer segments decrease over time and that there is no reduction in the loss of photoreceptor cells following therapy (Sarra et al 2001). Our investigations have excluded procedure-related damage, vector toxicity and immune responses as major factors which might counteract the benefits of ultrastructural improvements (Sarra et al 2001). There are a number of possible alternative explanations. We consider the major factors to be delayed onset or inappropriate levels of transgene expression or an insufficient transduction rate. A combination of all three and/or other factors may be important.

We wanted to determine whether we could improve photoreceptor survival as well as improving function by combining neuroprotection with gene replacement. Intraocular delivery of a variety of neurotrophic factors has been widely investigated as a potential treatment for retinal dystrophy. A number of studies have demonstrated the effect of vector-mediated ciliary neurotrophic factor (CNTF) gene expression on photoreceptor cell loss, suggesting that this may be an effective treatment for human retinal dystrophies. These studies have focused on animals in which there is very little function and concentrated on the morphology of the retina. A recent report, however, has evaluated long-term AAV-mediated CNTF gene expression in a transgenic mouse model of retinal disease caused by a dominant mutation in the *Prph2* gene, in which retinal function is relatively normal and only declines slowly with photoreceptor cell loss. Bok et al (2002) found a 23% preservation of photoreceptor cells as compared to the untreated side at 5 months after treatment, but observed reduced ERG recordings following treatment. In order to evaluate CNTF gene delivery as a potential treatment, we used the *Prph2*[Rd2/Rd2] mouse. CNTF was expressed intraocularly using AAV-mediated gene delivery either by itself or, in a second treatment group, combined with AAV-mediated gene replacement therapy of peripherin 2 (Schlichtenbrede et al 2003a,b). We confirmed in both groups of animals that CNTF reduces the loss of photoreceptor cells. Visual function, however, as assessed over a time course by ERG, was significantly reduced compared with untreated controls. Furthermore CNTF gene expression negated the effects on function of gene replacement therapy. In order to test whether this deleterious effect is only seen when degenerating retina is treated, we recorded ERGs from wild-type mice following intraocular injection of AAV expressing CNTF. Here a marked deleterious effect was noted, in which the b-wave amplitude was reduced by at least 50%.

The alteration in ERG trace may reflect the morphological changes caused by *CNTF* gene expression. The changes we have observed — a less well-ordered outer nuclear layer (ONL) and, in animals also treated with a gene replacement vector, less outer segment material — are consistent with recent reports that suggest that exposure to high levels of CNTF may have an impact on photoreceptor differentiation (Bok et al 2002, Caffe et al 2001). However, CNTF has been shown to exert an indirect effect on photoreceptor cell survival through cells of the inner retina and Müller cells. Wahlin et al (2001) have demonstrated that following intraocular injection of recombinant CNTF, signalling pathways are activated in Müller, ganglion and amacrine cells, but not photoreceptors. This effect was found both in normal and degenerating rodent retinae. It is the inner retinal neurons and Müller cells that are responsible for generating the b-wave. Thus the effect of CNTF on these cells may also be counteracting the potential benefit of photoreceptor preservation. Furthermore, CNTF overexpression might stimulate remodelling of the inner retina, leading to a change in cell function, a decrease in b-wave amplitude and subsequently to dedifferentiation of the photoreceptor cell, as reflected by the changes in chromatin staining (Bok et al 2002) It has been suggested that in the course of retinal degeneration, loss of input from photoreceptors leads to input-dependent secondary neurons of the inner retina 'seeking replacement' for the input of lost photoreceptors. It would appear that the remodelling of the inner retina that normally occurs in the course of degeneration is exacerbated by the expression of CNTF.

Our studies have shown that intraocular treatment with CNTF may result in unwanted side effects. These effects might be tolerable provided CNTF administration is only temporary. This might be achieved through the use of an inducible-promoter or a pharmacological slow release device with later removal of the ocular insert. To achieve a sustained morphological rescue without causing retinal damage a fine balance would have to be reached, requiring detailed dose–response studies, further complicated by the relatively high individual variability in the degenerating retina. Alternatively, other neurotrophic molecules should be evaluated for safety and efficacy. In retinal dystrophies, the degenerating retina undergoes a highly complex remodelling process, where any treatment-induced negative stimuli must be avoided.

To date most of the studies aimed at developing gene therapy for inherited retinal degenerations have focused on treatment of photoreceptor cell defects. Although we have shown that we can improve photoreceptor structure and function following gene replacement therapy in the *Prph2*[Rd2/Rd2] mouse, so far we have been unable to slow the loss of cells. It is perhaps not surprising that we have been unable to affect the degeneration in this animal model — it has a major cellular defect. The $Prph2^{Rd2/+}$ mouse on the other hand has a relatively mild phenotype of abnormal outer segments, and we are able to slow the degeneration by AAV-mediated delivery of peripherin 2.

Various forms of retinal degeneration are caused by RPE defects and there are a number of animal models of disease, including the $RPE65^{-/-}$ dog and the Royal College of Surgeons (RCS) rat. There are a number of reasons why RPE-specific defects may generally prove to be more amenable to treatment. The transduction efficiency of RPE is often much higher than that of photoreceptor cells and partial correction of RPE function may have significant impact on photoreceptor function and survival. Recently, Acland et al (2001) described highly effective treatment of $RPE65^{-/-}$ dogs with rAAV containing the $RPE65$ gene.

In the RCS rat, the phagocytosis of photoreceptor cell debris (mainly composed of shed outer segments) by the RPE is defective due to a 409 bp deletion in the receptor tyrosine kinase gene $Mertk$ (D'Cruz et al 2000). MERTK is involved in the recognition and binding of outer segment debris, possibly due to the appearance of phosphatidylserine in the outer leaflet of the plasma membrane of the shed disks. In the absence of functional MERTK the accumulation of outer segment material in the subretinal space results in the loss of photoreceptor cells and the degeneration of the retina. Photoreceptor cell loss is detectable histologically and electroretinographically from postnatal day 18 onward and subretinal accumulation of debris is apparent at that time. Depending on pigmentation of the animal the degeneration is virtually complete after 2 to 3 months; little electroretinographical activity is detectable and few photoreceptor cells remain. Recently, retinitis pigmentosa (RP) patients have been identified that have mutations in the $MERTK$ gene (Gal et al 2000). In these patients the disease is characterized by night blindness at a young age, progressing rapidly to a loss of peripheral vision until a small island of central vision remains.

Recently, Vollrath et al (2001) reported the effect on RCS rats of subretinal injection of recombinant adenovirus (rAd) containing the $Mertk$ gene under control of the CMV promoter. As is the case with Ad, adeno-associated virus (AAV) is able to transduce RPE cells with a high efficiency. In contrast with rAd vectors, transgene expression mediated by rAAV vectors is maintained over long periods; after injection into rodents, expression typically lasts throughout the lifetime of the animal. As there is also no notable decrease in rAAV transgene expression a year after treatment of non-human primates (Bennett et al 1999), it is likely transgene expression in humans will also persist for long periods of time. In order to develop a treatment for this form of RP that might be suitable for clinical application, we have evaluated gene therapy approaches in the RCS rat using rAAV vectors.

Our results show that following transduction of RPE cells in the RCS rat with rAAV expressing *Mertk* the phagocytotic function of these cells are restored and the accumulation of photoreceptor cell debris in the subretinal space is reduced (Smith et al 2003). The rate of photoreceptor cell degeneration that results from the deposition of debris is slowed after treatment, even though it is not completely prevented. Photoreceptor function is still detectable in treated eyes 9 weeks after injection, whereas untreated eyes at the same time point provide no recordable activity. ERG recordings from one-month old RCS rats are already lower than in wild-type animals, suggesting that damage to photoreceptors has occurred by the time treatment with AAV vectors may take effect. Although the late onset of AAV-mediated expression may have a major impact on the efficacy of treating RCS rats, this would not be a concern for the treatment of patients. Whereas retinal degeneration in the rats is complete by 3 months, in RP patients with mutations in the *MERTK* gene the disease process occurs over several decades. The incomplete rescue in this model might also be explained by the fact that not all the debris between the photoreceptors and RPE is cleared. Not only has debris accumulated by the time AAV-mediated *Mertk* expression occurs, the level of transgene expression may not be sufficient to clear the backlog. Furthermore, only part of the retina is transduced. We estimate that subretinal injections in the rat cover roughly 50% of the retina. Even if the photoreceptor cells in these areas of benefit from a reduction in debris, degenerating photoreceptors in other parts of the retina may have a negative impact on photoreceptor survival in treated areas of the retina. Despite the partial nature of the rescue, loss of function in the treated rats was slowed by several months. A similar approach in patients might be expected to slow the degeneration by years, rather than months, given the difference in the rates of degeneration in rats and humans. These results therefore provide strong support for developing AAV-based gene therapy approaches for RP patients with mutations in *MERTK*. We are now screening families for mutations in this gene in anticipation of a clinical trial.

Acknowledgements

I am very grateful to the Foundation Fighting Blindness, The Sir Jules Thorn Charitable Trust, The British Retinitis Pigmentosa Society and The Wellcome Trust for supporting our work on gene therapy for inherited retinal degeneration.

References

Acland GM, Aguirre GD, Ray J et al 2001 Gene therapy restores vision in a canine model of childhood blindness. Nat Genet 28:92–95

Ali RR, Sarra GM, Stephens C et al 2000 Restoration of photoreceptor ultrastructure and function in retinal degeneration slow mice by gene therapy. Nat Genet 25:306–310

Apfelstedt-Sylla E, Theischen M, Ruther K et al 1995 Extensive intrafamilial and interfamilial phenotypic variation among patients with autosomal dominant retinal dystrophy and mutations in the human RDS/peripherin gene. Br J Ophthalmol 79:28–34

Bainbridge JWB, Mistry A, Schlichtenbrede F et al 2003 Stable rAAV-mediated transduction of rod and cone photoreceptors in the canine retina. Gene Ther 10:1336–1344

Bennett J, Maguire AM, Cideciyan AV et al 1999 Stable transgene expression in rod photoreceptors after recombinant adeno-associated virus-mediated gene transfer to monkey retina. Proc Natl Acad Sci USA 96:9920–9925

Bessant DA, Ali RR, Bhattacharya SS 2001 Molecular genetics and prospects for therapy of the inherited retinal dystrophies. Curr Opin Genet Dev 11:307–316

Bok D, Yasumura D, Matthes MT et al 2002 Effects of adeno-associated virus-vectored ciliary neurotrophic factor on retinal structure and function in mice with a p216l rds/peripherin mutation. Exp Eye Res 74:719–735

Caffe AR, Soderpalm AK, Holmqvist I, van Veen T 2001 A combination of CNTF and BDNF rescues rd photoreceptors but changes rod differentiation in the presence of RPE in retinal explants. Invest Ophthalmol Vis Sci 42:275–282

D'Cruz PM, Yasumura D, Weir J et al 2000 Mutation of the receptor tyrosine kinase gene Mertk in the retinal dystrophic RCS rat. Hum Mol Genet 9:645–651

Dudus L, Anand V, Acland GM et al 1999 Persistent transgene product in retina, optic nerve and brain after intraocular injection of rAAV. Vision Res 39:2545–2553

Gal A, Li Y, Thompson D et al 2000 Mutations in MERTK, the human orthologue of the RCS rat retinal dystrophy gene, cause retinitis pigmentosa. Nat Genet 26:270–271

Liang FQ, Aleman TS, Dejneka NS et al 2001 Long-term protection of retinal structure but not function using RAAV.CNTF in animal models of retinitis pigmentosa. Mol Ther 4:461–472

McGee Sanftner LH, Abel H, Hauswirth WW, Flannery JG 2001 Glial cell line derived neurotrophic factor delays photoreceptor degeneration in a transgenic rat model of retinitis pigmentosa. Mol Ther 4:622–629

Sarra GM, Stephens C, de Alwis M et al 2001 Gene replacement therapy in the retinal degeneration slow (rds) mouse: the effect on retinal degeneration following partial transduction of the retina. Hum Mol Genet 10:2353–2361

Sarra GM, Stephens C, Schlichtenbrede FC et al 2002 Kinetics of transgene expression in mouse retina following sub-retinal injection of recombinant adeno-associated virus. Vision Res 42:541–549

Schlichtenbrede FC, MacNeil A, Bainbridge JWB et al 2003a Intraocular gene delivery of ciliary neurotrophic factor results in significant loss of retinal function in normal mice and in the Prph2Rd2/Rd2 model of retinal degeneration. Gene Ther 10:523–527

Schlichtenbrede FC, da Cruz L, Stephens C et al 2003b Long-term evaluation of retinal function in Prph2Rd2/Rd2 mice following AAV-mediated gene replacement therapy. J Gene Med 5:757–764

Smith AJ, Schlichtenbrede FC, Tschernutter M, Bainbridge JW, Thrasher AJ, Ali RR 2003 AAV-mediated gene transfer slows photoreceptor loss in the RCS rat model of retinitis pigmentosa. Mol Ther 8:188–195

Travis GH, Sutcliffe JG, Bok D 1991 The retinal degeneration slow (rds) gene product is a photoreceptor disc membrane-associated glycoprotein. Neuron 6:61–70

Vollrath D, Feng W, Duncan JL et al 2001 Correction of the retinal dystrophy phenotype of the RCS rat by viral gene transfer of Mertk. Proc Natl Acad Sci USA 98:12584–12589

Wahlin KJ, Adler R, Zack DJ, Campochiaro PA 2001 Neurotrophic signaling in normal and degenerating rodent retinas. Exp Eye Res 73:693–701

DISCUSSION

Hauswirth: In the *rds* gene replacement therapy, what level of *rds* expression are you achieving relative to wild-type levels in the animal model? It doesn't look like it is very high.

Ali: That's a good question. We are going to try to find out.

Hauswirth: This provides a potential explanation for what photoreceptors still die, because of a haploinsufficient effect.

Ali: It is possible, but the degeneration in the rds heterozygote is very slow.

Hauswirth: But this has 50% of normal levels and from your data it didn't look like you had anywhere near this level.

Ali: These are not simple experiments to do, but we are thinking about doing a quantitative RT-PCR to measure this.

Bird: How does the ERG compare between the treated and the heterozygote?

Ali: The heterozygote is clearly better.

Molday: Have you checked whether ROM1 expression is increased?

Ali: No. It would be difficult to have intact outer segments without increased ROM1.

Molday: No, outer segments can form in the absence of ROM1.

McInnes: In the ROM1 knockouts the outer segments are pretty healthy looking.

Travis: Would you say that in the injected area you are transducing close to all of the photoreceptors, or just a subset?

Ali: We are not transducing all the photoreceptor cells in the area. It is variable.

Travis: Having a situation where there are untransduced photoreceptors around a few transduced receptors expressing at a low level will be especially difficult to unravel.

Ali: It is a bit of an unusual situation because the *rds* mouse has no function. The key issue is whether we can improve the function and slow the degeneration. Just slowing the degeneration means very little in this context.

Travis: The fact that you are making outer segments is significant, though.

Ali: We are making progress.

Kaleko: With respect to the slow loss of the improvement in the rds model, is it possible that the promoter shut off over time or that there was an inflammatory response to the foreign transgene product in these animals?

Ali: In terms of inflammatory response, we don't really know, although I wouldn't absolutely rule it out, I think it is unlikely. We have carried out some experiments to show that we are not getting inactivation of the promoter. The inflammatory response is unlikely because we can get long-term expression of green fluorescent protein (GFP), although I accept your point that perhaps peripherin is more immunogenic than GFP.

Kaleko: Sometimes inflammatory responses can be fairly subtle. I have a second question. Do you think that the vector entry pathway into the photoreceptors is through the help of heparan sulfate glycosaminoglycans?

Hauswirth: For type 2 it definitely is.

Bok: Which isotype are you using?

Ali: Type 2.

Kaleko: Will there be a problem with subretinal injection under the macula?

Ali: I'm not the best person to answer this.

Sahel: Once you detach the macula the outer segments start to disappear. It may take many months before they regenerate. Did you detach the retina in your injections?

Ali: We do not cause a full detachment but we do detach between 40–60% and the ERG function seems to be normal.

Sahel: Certainly, the dog can elicit a lot of inflammatory responses by itself depending on the point of entry of the injection. It is unreliable.

Ali: Sometimes complete retinal detachments are performed when macular translocations are performed.

Sahel: I don't advise this.

Bird: I don't think the surgeon would see it as a problem. They are an adventurous group. But it is the case that surgical detachment of the macula with reattachment does not appear to be a very noxious exercise, as opposed to natural detachment where the subretinal space is open for a long time. If this is seen as a way of preserving function that would inevitably be lost, this will not be seen as a major problem.

Sahel: Steve Fisher (Rex et al 2002) has shown that once the retina is detached that there many markers of the inflammatory response present.

Ali: He keeps the retina artificially detached. We are carrying out a transitory detachment. It re-attaches within 24 hours. This type of detachment may not be as big a problem as one might first think. We have not found much evidence for damage.

Sieving: That is hedging the question. Why not do the experiment and publish the results? It is an important issue. It is hard for me to imagine that one can elevate the photoreceptors out of the nice RPE pockets that ensheathe the rod outer segments and then expect that the rods will slip right back into the sheaths afterwards. Might not the pigment epithelium sheaths collapse and then later on regrow around the outer segments? About 20 years ago I recorded an ERP (early receptor potential), in a patient several years after he underwent macula reattachment surgery. The ERP provides a measure of rhodopsin quantity in the rods and thereby gives information about rod outer segment length. Although this patient had 20:20 acuity, the ERP amplitude was only half normal, indicating that the macular photoreceptor outer segments were only half their normal length. In other words, the effect of his detachment persisted for many months. It is worth getting some data on this issue of structural retinal effects following artificial detachment and reattachment.

Sahel: There are two indirect ways to study this. First, in patients with retinal detachment, even if the macula is detached just for one day, it can take weeks or months to see total recovery. Second, if you look at the CTF to VCTF reattachment there is sometimes a tiny amount of liquid under the macula. Clinically it looks OK, but if you look carefully it is not that good.

Ali: It is obvious that if you have a normal healthy retina, the last thing we want to do is to detach it. If you have a severe disease process and there is no other way of doing anything then one might contemplate a retinal detachment.

Dryja: We just have to do the studies. One thing that surgeons might agree with is that repeated macular injections would not be good. It hard to imagine that one could do this every six months and expect the retina to last a lifetime.

Zack: Given the results that you have found, that a subretinal injection causes ganglion cell expression in the dog, how good is photoreceptor expression if you inject intravitreally? This might work for the macula and fovea in particular, since this is so thin.

Ali: That result is just because we are putting so much in and we are getting leakage out into the vitreous.

Zack: Have you tried injecting this much virus into the vitreous in the dog?

Ali: No.

Hauswirth: We have done the experiments in pigs and we only get ganglion cell transduction. The idea of doing it in the macula has some merit. Since there are many fewer cells that have to be passed through from the macula to get into photoreceptors, perhaps if you added a little heparin to compete with the initial binding at the proximal surface you could get penetration of a small viral vector.

Dryja: If it works, the retinal surgeons will figure out ways to administer the virus.

Aguirre: There seemed to be a great variability in the amount of rhodopsin and the outer segment length in the rds mouse that was treated. Is this a factor of time after injection?

Ali: Yes, you are right.

LaVail: In your *Mertk* experiments, it didn't look like there were many inner and outer segments there. From our experience in RCS rats, you are not going to get nearly the distribution that you get in any normal retina or other degenerate retinas because of all the debris in RCS rats. If this is the case, I was amazed at your ERG findings. You got demonstrable ERG b-waves with AAV RPE65.

Ali: But we had better results with the CMV.

LaVail: To have a relatively small rescue in a restricted area, and to have that much of an ERG response seems incompatible.

Ali: The results I showed were from repeated double injections, so we did four in all. To try to optimize our rescue we started to carry out multiple injections.

Aguirre: Are these at the same site?

Ali: No, different sites.

Reference

Rex TS, Fariss RN, Lewis GP, Linberg KA, Sokal I, Fisher SK 2002 A survey of molecular expression by photoreceptors after experimental retinal detachment. Invest Ophthalmol Vis Sci 43:1234–1247

General discussion I

Farber: Earlier I was talking to Paul Sieving and he described some interesting data on treating patients with Accutane (isotretinoin). Paul, could you describe these?

Sieving: Yes. We have tried an Accutane experiment with two subjects, to learn whether we could alter the rate of rhodopsin regeneration as measured by dark adaptation time course. Once we had found that in rodents that treatment with *cis*-retinoic acid was slowing rhodopsin degeneration, we tried this in people using dark adaptation psychophysics and also in monkeys by recording the electroretinogram (ERG). While I cannot be positive that we performed the experiment correctly in the two subjects, we did not find an effect on dark adaptation after 5 mg/kg Accutane. The only effect was that both subjects acquired substantial headaches that persisted for a couple of days. In the case of the monkeys, I believe that we did do the experiment correctly, as we applied the same technique as we had developed for mouse and rat. With 5 mg/kg subcutaneous dose in monkey, we did not observe any perceptible slowing of ERG recovery following a bleaching light exposure that could be attributed to a rhodopsin cycling effect.

I am curious whether this might work for Stargardt's: I would like to think it would. We had found that slowing of retinoid cycling and rhodopsin regeneration diminished the extent of retinal damage from acute light exposure in rats. We also studied the effect on retinal degeneration in the RCS rat and in Matt LaVail's Pro23 rhodopsin rat, but we did not see a generalized protection in either case that extended beyond light-damage protection. We then thought that other genetic models in which rhodopsin cycling was critical might be appropriate targets, such as *Abcr4*. With Gabe Travis there are now preliminary data that look really hopeful on this model.

We were also curious about whether Accutane was the only retinoid compound that had an effect on retinoid cycling. Part of our motivation was that we were in the process of filing a patent application at the University of Michigan for this idea in 2000. We tried a large range of retinoid analogues, and we found that many of them had a comparable effect on the ERG recovery from bleaching light, which was interpreted as a slowing of rhodopsin regeneration and retinoid cycling. Does the effect of retinal light compounds extend beyond Accutane alone? Does anyone know of other diseases that might involve retinoid cycling?

Bok: Certainly, *RPE65* mutation is an example where reduced retinoid can be beneficial under certain circumstances. The C2J mouse is an example where less retinoid can be better.

Sieving: Has the *Abcr4* mutant been crossed onto one of these backgrounds?

Bok: Gabriel Travis has looked at his *Abcr4* knockouts in the context of whether C2J is present or not.

Travis: We moved it onto an albino background, and we wanted to make sure we had the so-called wild-type RPE65, the sensitizing mutation. We do.

Thompson: Another disease involving retinoid cycling would be retinal degeneration resulting from mutations in the *LRAT* gene. Mutations in *LRAT* would be expected to result in decreased levels of 11-*cis*-retinal, as in the case of mutations in *RPE65*. On the other hand, Jack Saari suggested to me that *LRAT* mutations might also result in high concentrations of free vitamin A in the retina and retinal pigment epithelium (RPE), and that this could produce toxic effects. This is what he felt he was seeing in his own work with the CRBP knockout mouse. If it does turn out that absence of 11-*cis*-retinal is the main problem resulting from *LRAT* mutations, then strategies that work for *RPE65* mutations may work equally well for patients with mutations in *LRAT*. However, mutations in *LRAT* appear to occur with much lower frequency than mutations in *RPE65*, and therefore the number of *LRAT* patients available for treatment could be low.

Bird: We talked to dermatologists and they took the view that it would be bad to give Accutane to people over a long period, just because of the side effects. Will the other compounds known to interfere with vitamin A metabolism inevitably have the same side effects?

Sieving: I do not have an answer for that.

Bird: I would say that any condition where there is elevated autofluorescence at the level of the pigment epithelium might conceivably be a disorder that would benefit from such an approach. This would include Stargardt disease, the cone dystrophies, the macular disorders associated with retinitis pigmentosa (RP), and even geographic atrophy with ageing. In all of these there is increased autofluorescence that accompanies cell death. It might have a much wider application than Stargardt disease.

Range of retinal diseases potentially treatable by AAV-vectored gene therapy

William W. Hauswirth*†‡, Quihong Li†, Brian Raisler*, Adrian M. Timmers†, Kenneth I. Berns*‡, John G. Flannery§, Matthew M. LaVail¶ and Alfred S. Lewin*‡

*Departments of *Molecular Genetics and †Ophthalmology, and ‡Powell Gene Therapy Center, University of Florida, Gainesville FL, §Departments of Vision Science and Neuroscience Group, University of California, Berkeley, CA, and ¶Departments of Anatomy and Ophthalmology, Beckman Vision Center, University of California, San Francisco, CA, USA*

Abstract. Viable strategies for retinal gene therapy must be designed to cope with the genetic nature of the disease and/or the primary pathologic process responsible for retinal malfunction. For dominant gene defects the aim must be to destroy the presumably toxic gene product, for recessive gene defects the direct approach aims to provide a wild-type copy of the gene to the affected retinal cell type, and for diseases of either complex or unknown genetic origin, more general cell survival strategies that deal with preserving affected retinal cells are often the best and only option. Hence examples of each type of therapy will be briefly discussed in several animal models, including ribozyme therapy for autosomal dominant retinitis pigmentosa in the transgenic P23H opsin rat, β-PDE gene augmentation therapy for autosomal recessive retinitis pigmentosa in the rd mouse, glial cell-derived neurotrophic factor (GDNF) gene therapy for autosomal dominant RP in the transgenic S334ter opsin rat and pigment epithelial cell-derived neurotrophic factor (PEDF) gene therapy for neovascular retinal disease in rodents. Each employs a recombinant AAV vectored passenger gene controlled by one of several promoters supporting either photoreceptor-specific expression or more general retinal cell expression depending on the therapeutic requirements.

2004 Genetics to gene therapy of retinal dystrophies. Wiley, Chichester (Novartis Foundation Symposium 248) p 179–194

Gene therapy strategies for retinal diseases can, in the most general sense, be divided into three classes. (1) For dominant gene defects, the aim must be to destroy the presumably toxic gene product either at the gene, mRNA or protein level. As is often the case with dominant gene defects the remaining wild-type gene copy is insufficient for normal cellular function, haploinsufficiency. In this case, in addition to removing the mutant gene or gene product extra wild-type gene copies must also be provided. (2) For recessive gene defects, the approach must involve restoring the 'missing' gene function, usually through direct delivery and expression of wild-type copies of the non-functional gene. (3) Finally, for either dominant or recessive defects, or diseases of either complex genetic aetiology or of unknown genetic origin, more general 'survival' gene strategies are often the only option. This last option is often the most attractive, at least in theory, because it is largely independent of the genetic basis of the disease and one treatment modality may be useful for several forms of retinal maladies.

Therapeutic approaches employing AAV-vectored gene delivery and expression in retinal cells have shown great promise recently in animals. Here we emphasize the range of retinal diseases that are potentially amenable to this general strategy by briefly describing our collaborative work using each class of therapy in several animal models which taken together exhibit a broad spectrum of retinal diseases. More specifically, we summarize progress on ribozyme treatment for autosomal dominant retinitis pigmentosa (RP) in the P23H opsin rat and gene augmentation for autosomal recessive RP in the *rd* mouse. In addition, progress on more general cell survival strategies is also summarized employing the gene for glial cell-derived neurotrophic factor (GDNF) in a dominant animal model of RP, the transgenic S334-ter opsin rat, and pigment epithelial cell-derived factor (PEDF) for retinal neovascularization in the ischaemic neonatal mouse. Each approach employs a recombinant AAV vectored passenger gene controlled by one of several promoters supporting either photoreceptor-specific expression or more general retinal cell expression depending on the specific therapeutic requirement. Each is a collaborative effort among several laboratories. Taken together, it is becoming clear that recombinant AAV vectors are versatile gene delivery vehicles that, in the context of the appropriate promoter and intraocular site of administration, have clear potential in a wide variety of retinal maladies.

Ribozyme therapy for autosomal dominant RP

Disease caused by dominantly inherited mutations typically involve production of a toxic gene product that alters cellular metabolism and/or leads to cell death. In the retina, numerous examples of photoreceptor degeneration caused by such mutant alleles can be found, particularly in the multiple genetic forms of autosomal dominant RP (ADRP). One promising therapeutic approach to such disorders is the use of ribozymes, small RNA molecules that cleave mutant transcripts in an allele-specific manner while leaving wild-type transcripts intact (Hauswirth & Lewin 2000). Because of their ability to act catalytically to cleave multiple substrate RNA molecules, ribozymes can be designed, tested and optimized kinetically *in vitro* to efficiently target dominant negative alleles causing ADRP (Shaw et al 2001). Expression of ribozymes has been shown to inhibit gene expression *in vitro* (Birikh et al 1997, Good et al 1997) and *in vivo* (Abounader et al 1999, Chowrira et al 1994, L'Huillier et al 1996, Lieber & Kay 1996, Macejak et al 1999) also suggesting their utility as treatment for ADRP.

AAV vectors have been designed to efficiently target *in vivo* expression of passenger genes to rodent rod photoreceptors using a limited portion of the rodent or bovine rod opsin promoter (Flannery et al 1997). Using this vector with a ribozyme gene designed against the mutant P23H (histidine substituted for proline at position 23) murine rod-opsin gene in transgenic rats (Lewin et al 1998), we demonstrated that ribozymes are effective *in vivo* therapeutics against dominant negative alleles. This rat is a model for the most common form of ADRP in the USA (Dryja et al 1991). P23H opsin ribozymes delivered to rods by AAV not only reduced mutant transcript levels *in vitro* (Drenser et al 1998), but also *in vivo* (Lewin et al 1998). In the transgenic rat retina, this resulted in a significant slowing of the progression of photoreceptor degeneration. In these experiments, two forms of ribozyme, hairpin and hammerhead ribozymes, were designed to target the mutant transcript, and both worked well. Vectors were injected subretinally at postnatal day 15 (P15), before significant photoreceptor degeneration, and the eyes examined between P60–P90. In this age window, uninjected control eyes in transgenic rats expressing the dominant P23H opsin allele had on average approximately 60% of the number of photoreceptors in age-matched, wild type retinas. In contrast, in the contralateral ribozyme treated eyes, 83–88% of the normal photoreceptor number remained ($P < 0.001$).

Electroretinographic (ERG) responses from a number of P23H rats were determined to assess whether AAV-ribozyme treatment also preserved the retina's physiological response to input light (Lewin et al 1998). Since one eye of an animal was typically treated with an AAV-ribozyme while the other served as a control, simultaneous recordings from both eyes eliminates animal-to-animal and flash-to-flash variation in ERG recordings. In non-transgenic normal rats, both a- and b-wave ERG amplitudes are within 5% of each other. Treatment by either active hammerhead or hairpin ribozymes resulted in scotopic b-waves 30–65% greater than in the contralateral control eyes, a highly significant improvement ($P < 0.005$). Inactive hammerhead or hairpin ribozymes also resulted in larger b-wave amplitudes than their respective control eyes, but only about 25% greater than controls, suggesting a reproducible, but lower level of rescue than for their catalytically active counterparts. This effect is most likely due to an antisense effect mediated by the ribozymes' short targeting arms on the ribozymes designed to anneal to the mutant mRNA. ERG amplitudes for PBS- and AAV-BOPS-*gfp* injected eyes were not significantly different from uninjected eyes. Rod a-wave amplitudes were as much as 93% greater than uninjected control partner eyes. In contrast, the a-wave amplitudes of neither PBS- nor AAV-BOPS-*gfp* treated eyes were different than controls. In summary, preservation of the retina's ability to respond physiologically to light demonstrates that appropriately designed and expressed ribozymes can lead to functional as well as structural photoreceptor rescue in a retina with a dominant negative disease.

Two important questions were left unanswered in this initial study: how permanent is the rescue and how late in the disease can ribozymes be introduced and still have a therapeutic effect? It seemed likely that ribozymes would persist in photoreceptors for an extended period because GFP reporter gene expression using the same vector–promoter combination persists in rats for at least 30 months (W. W. Hauswirth, J. G. Flannery, unpublished work). Additionally, although AAV elicits a modest humoral immune response in rodents and primates, AAV vectors do not lead to an inhibitory cellular immune response (Bennett et al 1999, Hernandez et al 1999). Whether delaying the time of ribozyme delivery alters its effectiveness addresses whether late stages of retinal degeneration will respond as well as early stages, a key question in human therapy where patients often present with measurable symptoms fairly late in the course of degeneration.

These issues were examined in the P23H rat using the initially validated ribozymes described above (LaVail et al 2000). For the long-term study of ribozyme therapy, the vector was administered at P15 and retinal structure and function analysed as late as P240. For analysis of later stage ribozyme gene transfer, vector was delivered at either P30 or P45, when 40–45% of the P23H rat photoreceptors had already been lost. When injected at P15, expression of either ribozyme markedly slowed the rate of photoreceptor degeneration for at least 8 months relative to the untreated partner retina (18 vs. 8 μm outer nuclear layer [ONL] thickness) and resulted in significantly greater ERG amplitudes ($P < 0.005$) at least up to P180. When injected at P30 or P45, nearly the same number of photoreceptors survived at P130 as seen when treatment was initiated at P15. Therefore, well-validated ribozymes designed against autosomal dominant alleles appear to be an effective, long-term therapy for ADRP and the treatment remains effective even when the gene transfer is done after significant photoreceptor cell loss.

Gene augmentation therapy for recessive retinal disease in the *rd* mouse

The *rd* mouse containing a homozygous mutation in the gene coding for the β subunit of phosphodiesterase (β-PDE) has long been used as an animal model for recessive RP. However the fast progression of the degeneration process (nearly all rods lost by P18) renders it a difficult model to use for experimentally evaluating virally vectored gene therapies. A transgenic mouse containing one wild-type copy of the β-PDE gene on the homozygous *rd* background (*rd/rd/tg*⁺) was made several years ago and reported to exhibit improved photoreceptor survival (Lem et al 1992). Transgenic *rd/rd/tg*⁺ mice were recovered from frozen embryos, and their retinal structure and function characterized by full field scotopic ERG and histologic analysis for up to 18 weeks after birth. Retinas of $rd/rd/tg^+$ mice have normal morphology and scotopic ERG responses up to 6–8 weeks of age. A progressive degeneration then begins, with a reduction of the outer nuclear layer to 1–2 rows of nuclei or less by 14–16 weeks. Thus, although retinas remained apparently normal until about P50, by P100 only about one row of the normal ten rows of photoreceptor nuclei remained. Because this time course of rod loss is much more amenable to gene therapy evaluation than the original rd mouse, we used it to test AAV-mediated β-PDE gene therapy to prevent or slow the degeneration process.

For therapeutic intervention, a wild-type murine β-PDE cDNA under control of a mouse rod opsin proximal promoter was cloned into an AAV and injected into the subretinal space of $rd/rd/tg^+$ mice at 3–4 weeks of age. Treated mice were followed by ERG and histology for up to 12 weeks post-injection. Retinas injected with AAV-β-PDE showed significant ($P < 0.01$) retinal preservation as measured by full-field scotopic ERG response 6, 8 and 10 weeks after injection. This correlated with histological preservation of the photoreceptor layer (2–4 more rows of nuclei than in untreated controls). Some of the treated animals had significant full field scotopic ERG response 12 weeks after injection ($\sim 100\,\mu$V max b-wave) while all uninjected eyes lost recordable scotopic ERG responses completely by this age. In no animals, however, did β-PDE gene augmentation preserve function for longer than 10 weeks post-injection, i.e. by P120 all treated retinas had only one or less rows of photoreceptor nuclei. This suggests that the apparently recessive rd allele may have a weakly dominant negative effect, a phenomenon that may not be uncommon in recessive disorders. In summary, a wild-type β-PDE gene controlled by a mouse opsin promoter and delivered via an AAV vector reproducibly slowed retinal degeneration in $rd/rd/tg^+$ mice, validating the feasibility of AAV-mediated gene augmentation therapy for slowing the progression of PDE-based retinal degenerative diseases, an important subclass of human RP.

Not all recessive retinal diseases are as refractory to sustained long-term gene augmentation therapy as the rd mouse. In a recently published study (Acland et al 2001, Bennett 2001, this volume), AAV-vectored RPE65 cDNA therapy in the recessive RPE65 mutant dog was more permanent and dramatic. This animal suffers severe night blindness at birth reflected by a 5–6 log elevation of ERG thresholds, severely attenuated b-wave amplitudes and lack of detectable a-waves. The RPE65 dog is considered a model for RPE65 human Leber congenital amaurosis, a severe congenital form of childhood blindness. Since the defect is in an RPE-specific gene, we employed the CBA promoter (fusion of the chicken β-actin promoter and CMV immediate early enhancer) that supports strong expression in both the RPE and photoreceptors when injected into the subretinal space. Visual function was rescued up to at least 9 months after AAV-RPE65 treatment. At 3–4 months post-injection, ERG b-wave thresholds were improved by about 4 log units, ERG a- and b-wave amplitudes were restored, and waveforms were similar to normals. Cortical evoked potentials, pupillary responses and behavioural assessment supported the ERG evidence of restored visual function. Thus, gene augmentation therapy mediated by AAV for RPE genes appears feasible and can be of long duration.

GDNF therapy in the transgenic S334ter rat model of autosomal dominant RP

Glial cell-derived neurotrophic factor (GDNF) is a member of the transforming growth factor-β superfamily, and has been shown to have broad neuroprotective effects (Choi-Lundberg et al 1997, Gash et al 1996, Kaddis et al 1996, Matheson et al 1997, Oppenheim et al 1995). As an injected protein, GDNF protects photoreceptors histologically and functionally in the *rd* mouse (Frasson et al 1999). We therefore evaluated the therapeutic potential of GDNF to promote photoreceptor survival in the transgenic S334ter rod-opsin rat, a model for dominant RP.

Based on our experience with other neurotrophins (Bok 2003, this volume), we used the CBA promoter (see above) to express GDNF from the AAV vector because expression occurs in both photoreceptors and RPE when the vector is injected into the subretinal space. By Western blot analysis, expression of human recombinant GDNF was detected in retinas transduced with AAV-GDNF, but not from untransduced or AAV-GFP-transduced retinas. GDNF localized to both photoreceptor inner and outer segments and to RPE cells. At P60, 45 days post-treatment, both superior and inferior hemispheres of AAV-GDNF treated retinas exhibited increased ONL thickness compared to uninjected or AAV-GFP treatments. In superior hemispheres, GDNF-treated retinas had a significantly ($P < 0.05$) greater average ONL thickness of 23.8+4.1μm than AAV-GFP controls (16.3+2.5 μm) or uninjected eyes (15.4+2.2 μm). In inferior hemispheres, GDNF-treated retinas had an average ONL thickness of 28.8+2.8 μm compared to 21.9+3.0 μm in AAV-GFP controls or 21.4+2.6 μm in uninjected ($P < 0.05$). Six to eight rows of ONL photoreceptor nuclei were present in the superior retinas treated with AAV-GDNF compared to 4–6 rows in AAV-GFP treated retinal and 2–3 rows in uninjected controls.

ERG analysis showed functional preservation paralleling structural rescue. At P60, AAV-GDNF eyes showed a significant ($P < 0.05$) retention of the mean a-wave amplitude, 74.3+15.5 μV compared to 60.9+11.3 μV in AAV-GFP control and 53.0+9.5 μV in uninjected. AAV-GDNF treated retinas also had increased b-wave mean amplitude, 360.9+50.8 μV compared to 315.5+42.2 μV in AAV-GFP controls or $298.5+33.2\,\mu V$ in uninjected ($P < 0.05$). Neither neovascularization, inflammation, nor alterations in photoreceptor nuclear morphology were observed in eyes treated with AAV-CBA-GDNF. Therefore, GDNF has potential as a gene therapy for photoreceptor degenerations apparently without the side effects noted with CNTF (Bok 2003, this volume).

PEDF therapy in the ischaemic neonatal mouse model of retinal neovascularization

The formation of new blood vessels (angiogenesis) in the mature retinal vasculature and its extension into the retina proper is the central factor leading to vision loss in diseases such as retinopathy of prematurity (ROP) and proliferative diabetic retinopathy (PDR), the most frequent cause of blindness among the working-age population (D'Amore 1994). Unfortunately no treatment demonstrates long-term success for patients. Pigment epithelium derived factor (PEDF), first purified from human retinal pigment epithelial cultures as a factor that induces neuronal differentiation of cultured retinoblastoma cells (Shweiki et al 1992, Steele et al 1993), has been recently shown to regulate normal angiogenesis in the eye (Dawson et al 1999). PEDF is down-regulated by hypoxia and induced in the retina as a result of hyperoxia; it is a very potent inhibitor of corneal neovascularization (NV) and prevents endothelial cell migration towards a wide variety of angiogenic inducers (Dawson et al 1999). PEDF therefore appears to be a major angiogenic regulator of the retinal vasculature and is an excellent candidate gene for therapy aimed at limiting ocular NV. Recently, as a systemically injected protein, PEDF inhibited retinal neovascularization (NV) in an ischaemic neonatal mouse model of ROP (Stellmach et al 2001). We therefore tested whether intraocular AAV-delivered PEDF cDNA when expressed from the CBA promoter would reduce retinal NV in this same model.

AAV-CBA-PEDF vector was delivered as a single intravitreal injection to one eye of P1 mouse pups. At P7, mouse pups with nursing dams were exposed to 72% oxygen for 5 days and returned to room air on P12. This induces retinal NV that is maximal at P17. The level of NV is then quantified by counting the number of vascular endothelial cell nuclei internal to the inner limiting membrane in each section across the retina. At P17 the level of PEDF in vector-injected eyes ranged from 20–76 ng per eye, well above the threshold level estimated from systemic PEDF protein injections (Stellmach et al 2001) of about 5 ng per eye. Upon analysis of retinal NV in eyes injected with rAAV-CBA-PEDF, there was a decrease in the number of neovascular endothelial cell nuclei compared with the contralateral control eye by 35% on average (Raisler et al 2002). This suggests that

AAV-mediated PEDF treatment has been effective in limiting retinal NV in this model.

Conclusion

Given that it is now well established that AAV-vectored genes can exhibit high expression levels for prolonged periods of time in retinal cells *in vivo*, the critical elements for a therapeutic reagent include: (1) a gene appropriate for the disease, validated *in vitro* if possible, (2) a promoter allowing proper retinal cell type regulation and expression level of the therapeutic gene and (3) delivery of the AAV vector to the appropriate intraocular site. When these three criteria are met, it is becoming clear that AAV-mediated expression of genes in the retina will have broad therapeutic potential for a wide range of retinal diseases.

Acknowledgements

Research support was provided by the NIH (EY01919, EY06842, EY11123, NS36302, EY11596, EY13101), Juvenile Diabetes Foundation, Macula Vision Research Foundation, Research to Prevent Blindness, Inc., That Man May See and Foundation Fighting Blindness.

References

Abounader R, Ranganathan S, Lal B et al 1999 Reversion of human glioblastoma malignancy by U1 small nuclear RNA/ribozyme targeting of scatter factor/hepatocyte growth factor and c-met expression. J Natl Cancer Inst 91:1548–1556

Acland GM, Aguirre GD, Ray J et al 2001 Gene therapy restores vision in a canine model of childhood blindness. Nat Genet 28:92–95

Bennett J 2003 Gene therapy for Leber congenital amaurosis. In: Retinal dystrophies: functional genomics to gene therapy. Wiley, Chichester (Novartis Found Symp 255), p 195–207

Bennett J, Maguire AM, Cideciyan AV et al 1999 Stable transgene expression in rod photoreceptors after recombinant adeno-associated virus-mediated gene transfer to monkey retina. Proc Natl Acad Sci USA 96:9920–9925

Birikh KR, Heaton PA, Eckstein F 1997 The structure, function and application of the hammerhead ribozyme. Eur J Biochem 245:1–16

Bok D 2003 Gene therapy of retinal dystrophies: achievements, challenges and prospects. In: Retinal dystrophies: functional genomics to gene therapy. Wiley, Chichester (Novartis Found Symp 255), p 4–16

Choi-Lundberg DL, Lin Q, Chang YN et al 1997 Dopaminergic neurons protected from degeneration by GDNF gene therapy. Science 275:838–841

Chowrira BM, Pavco PA, McSwiggen JA 1994 In vitro and in vivo comparison of hammerhead, hairpin and hepatitis delta virus self-processing ribozyme cassettes. J Biol Chem 269:25856–25864

D'Amore PA 1994 Mechanisms of retinal and choroidal neovascularization. Invest Ophthalmol Vis Sci 35:3974–3979

Dawson DW, Volpert OV, Gillis P et al 1999 Pigment epithelium-derived factor: a potent inhibitor of angiogenesis. Science 285:245–248

Drenser KA, Timmers AM, Hauswirth WW, Lewin AS 1998 Ribozyme-targeted destruction of RNAs associated with autosomal-dominant retinitis pigmentosa. Invest Ophthalmol Vis Sci 39:681–689

Dryja TP, Hahn LB, Cowley GS, McGee TL, Berson EL 1991 Mutation spectrum of the rhodopsin gene among patients with autosomal dominant retinitis pigmentosa. Proc Natl Acad Sci USA 88:9370–9374

Flannery JG, Zolotukhin S, Vaquero MI, LaVail MM, Muzyczka N, Hauswirth WW 1997 Efficient photoreceptor-targeted gene expression in vivo by recombinant adeno-associated virus. Proc Natl Acad Sci USA 94:6916–6921

Frasson M, Picaud S, Leveillard T et al 1999 Glial cell line-derived neurotrophic factor induces histologic and functional protection of rod photoreceptors in the rd/rd mouse. Invest Ophthalmol Vis Sci 40:2724–2734

Gash DM, Zhang Z, Ovadia A et al 1996 Functional recovery in parkinsonian monkeys treated with GDNF. Nature 380:252–255

Good PD, Krikos AJ, Li SX et al 1997 Expression of small, therapeutic RNAs in human cell nuclei. Gene Ther 4:45–54

Hauswirth WW, Lewin AS 2000 Ribozyme uses in retinal gene therapy. Prog Retin Eye Res 19:689–710

Hernandez YJ, Wang J, Kearns WG, Loiler S, Poirier A, Flotte TR 1999 Latent adeno-associated virus infection elicits humoral but not cell-mediated immune responses in a nonhuman primate model. J Virol 73:8549–8558

Kaddis FG, Zawada WM, Schaack J, Freed CR 1996 Conditioned medium from aged monkey fibroblasts stably expressing GDNF and BDNF improves survival of embryonic dopamine neurons in vitro. Cell Tissue Res 286:241–247

L'Huillier P, Soulier S, Stinnakre MG et al 1996 Efficient and specific ribozyme-mediated reduction of bovine α-lactalbumin expression in double transgenic mice. Proc Natl Acad Sci USA 93:6698–6703

LaVail MM, Yasumura D, Matthes MT et al 2000 Ribozyme rescue of photoreceptor cells in P23H transgenic rats: long-term survival and late-stage therapy. Proc Natl Acad Sci USA 97:11488–11493

Lem J, Flannery JG, Li T, Applebury ML, Farber DB, Simon MI 1992 Retinal degeneration is rescued in transgenic rd mice by expression of the cGMP phosphodiesterase β subunit. Proc Natl Acad Sci USA 89:4422–4426

Lewin AS, Drenser KA, Hauswirth WW et al 1998 Ribozyme rescue of photoreceptor cells in a transgenic rat model of autosomal dominant retinitis pigmentosa. Nat Med 4:967–971

Lieber A, Kay MA 1996 Adenovirus-mediated expression of ribozymes in mice. J Virol 70:3153–3158

Macejak DG, Lin H, Webb S et al 1999 Adenovirus-mediated expression of a ribozyme to c-myb mRNA inhibits smooth muscle cell proliferation and neointima formation in vivo. J Virol 73:7745–7751

Matheson CR, Wang J, Collins FD, Yan Q 1997 Long-term survival effects of GDNF on neonatal rat facial motoneurons after axotomy. Neuroreport 8:1739–1742

Oppenheim RW, Houenou LJ, Johnson JE et al 1995 Developing motor neurons rescued from programmed and axotomy-induced cell death by GDNF. Nature 373:344–346

Raisler BJ, Berns KI, Grant MB, Beliqev D, Hauswirth WW 2002 Adeno-associated virus type-2 expression of pigmented epithelium-derived factor or Kringles 1-3 of angiostatin reduce retinal neovascularization. Proc Natl Acad Sci USA 99:8809–8814

Shaw LC, Skold A, Wong F, Petters R, Hauswirth WW, Lewin AS 2001 An allele-specific hammerhead ribozyme gene therapy for a porcine model of autosomal dominant retinitis pigmentosa. Mol Vis 7:6–13

Shweiki D, Itin A, Soffer D, Keshet E 1992 Vascular endothelial growth factor induced by hypoxia may mediate hypoxia-initiated angiogenesis. Nature 359:843–845

Steele FR, Chader G J, Johnson LV, Tombran-Tink J 1993 Pigment epithelium-derived factor: neurotrophic activity and identification as a member of the serine protease inhibitor gene family. Proc Natl Acad Sci USA 90:1526–1530

Stellmach VV, Crawford SE, Zhou W, Bouck N 2001 Prevention of ischemia-induced retinopathy by the natural ocular antiangiogenic agent pigment epithelium-derived factor. Proc Natl Acad Sci USA 98:2593–2597

DISCUSSION

McInnes: In the P23 mouse, why do the cells continue to die? Is the level of expression decreasing, or are those cells that weren't transfected?

Hauswirth: It appears that we are rescuing a certain fraction of cells. If we look at it again, you can take about 35% of the nuclei and add that to every time point on the baseline. The cells we have treated apparently stay treated, and those that have not received the treatment continue to die with the same kinetics.

LaVail: Keep in mind that a rat eye continues to thin because of the stretching of the retina around the curve of the eye. In wild-type the curves are parallel, as Bill Hauswirth has mentioned. It appears that the ribozymes have stopped the cell death altogether at a certain point, and this is maintained.

McInnes: Have you actually shown that it is the cells that weren't infected that die?

Hauswirth: That is just a supposition since *in situ* analysis for ribozymes is very difficult. This suggests that if we were to adopt Robin Ali's approach of multiple injections we would get more cells protected. We do know that we are reducing the levels of the mutant alleles in the treated eyes by about 20%.

Kaleko: You have an interesting decision to make. With AAV you have the choice of an intravitreal injection, which will give you a more global distribution, and which is significantly easier to achieve in the clinic than a subretinal injection. However, it will result in expression from ganglion cells, which could lead to retroaxonal transport of your protein into the brain. This may concern the FDA. Or you have the optional of a subretinal injection, which with the appropriate promoter could limit expression to the RPE. This is more difficult to do, but perhaps safer from an FDA perspective. Which way will you go?

Hauswirth: It is going to depend on the disease. There is actually a third possibility. Dr Peter Campochiaro has been able to show that it is possible to infect the retina by injecting a vector extra-ocularly. He can get the entire orbit expressing β galactosidase. There may therefore be on the horizon a very easy non-pathogenic way to deliver secretable agents.

Kaleko: I have seen those data. Currently, the expression levels in the eye are relatively low and seem to be dependent on the specific therapeutic protein.

Hauswirth: If we take that option off the table, my preference for a secretable agent is currently by delivering a vector to the vitreous, but we will have to be careful to measure any attendant potential toxicities.

Zack: A fourth possibility might be using Müller cells which secrete a lot of factors anyway. Various viruses are pretty good with Müller cells.

Hauswirth: We are trying to create an entire tool box of AAV vectors that will work in specific cell types, and Müller cells are one of these.

Kaleko: Can you briefly review the data indicating that PEDF works through a blockade of VEGF. I have not been able to verify this.

Hauswirth: We can't nail this down either. They are clearly reciprocally regulated. When the vasculature is developing there is lots of VEGF but at the end of vascularization when we want to inhibit further vascularization PEDF comes up and VEGF goes down.

Kaleko: We have the same challenge that you do: to choose a transgene. A transgene with a known mechanism will have more predictive value from the animal models to humans. If we can nail down the mechanism for PEDF it would make it a more palatable transgene.

Hauswirth: There are more obvious mechanisms. Robin Ali (among others) has looked at the vectored expression of the soluble Flt receptor so you are competing for VEGF binding by a soluble receptor. This seems to be working and is much easier to understand since the mechanism is a simple competition with VEGF for its cellular receptor.

Ali: Have you tried PEDF driven by the CMV promoter?

Hauswirth: Yes. In both cases I called the expression level 'fair'. It was still above the therapeutic threshold, but not anywhere as near above it as CBA.

Ali: Have you looked at it in terms of affecting the vascularization?

Hauswirth: No.

Ali: We have run this through our model and we don't see an effect of CMV-PEDF.

Hauswirth: Perhaps it is too close to the threshold.

Swaroop: You have very nicely described various challenges, including delivery and target specificity. What about the regulation in terms of gene placement? Over-expression or mis-expression in the same cell can also cause various problems. Are you thinking about using natural promoters to examine this? Aren't you worried about this possibility?

Hauswirth: Yes. There are a lot of these tough decisions that we are going to have to make very shortly. We know that the tetracycline-doxocycline promoter system can effectively regulate expression of our genes. The problem we have is that AAV is small and these promoter systems take up so much space that we can't use them in a single vector. Instead we have to use very simple promoters, and I still have my doubts whether these will give us enough expression for therapy.

Swaroop: Sometimes very small sequences of 200–300 bp are enough to give cell type-specific expression. This ought to be considered. If you use an inducible promoter you are not regulating the amount of RNA that is being produced. This could be deleterious. It might be better to use a natural promoter.

Hauswirth: I don't get your point: how is that adding to the safety profile? I can express less right now by using less vector. I don't see how you are getting a level of modulation.

Swaroop: If you use a natural promoter it will give a natural level of expression.

Hauswirth: I don't know whether that is true, and it may still not be simple: we have an example using the RPE65 promoter itself. Large and small RPE65 promoters express very little passenger gene in the RPE, but a specific size works pretty well to maintain RPE specificity. However, if we have too long or too short a promoter we get poor expression. Even the best RPE65 promoter so far is much weaker than the CBA promoter, but it may be sufficient for therapy, and we are trying now to test this idea.

LaVail: Marty Friedlander, I gather you have had some experience with anti-angiogenic approaches. Do you have anything to add here?

Friedlander: I was going to ask a question about selectivity in targeting. I know that Bill Hauswirth has chosen AAV, but there is a lot of work going on with developing pseudotyped adenovirus vectors. There are 50 serotypes many of which have different receptors. So what about the concept of getting the transgene to specific cell types on the basis of receptor binding or internalization mechanisms? Have you heard anything about that?

Hauswirth: It is possible to do the same things with AAV now. There are several locales in capsid proteins where it is now possible to insert ligands. It has been shown that even AAV type 2 can be targeted to cells that would normally be refractory, by targeting for example the ApoE ligand. You might argue that we are trying to make AAV into an adenovirus, or vice versa.

Friedlander: Another concept is perhaps using viral vectors to selectively transfect endothelial progenitor cells. You can isolate sub-populations of haematopoietic stem cells that contain endothelial progenitor cells and transfect them with your favourite transgene. These stem cells can then be re-injected where they will target a specific tissue (e.g. neovasculature) and produce the transgene. The effects of these transgene products produced locally can then be assessed. This may be another way to start thinking about getting transgenes to work.

Hauswirth: That is an incredibly powerful approach. It has a lot of merit.

LaVail: Speaking of proliferation, in your ROP model where you are getting this pulse of expression of PEDF, why isn't that maintained?

Hauswirth: I think it reflects the relatively short half-life of PEDF.

LaVail: Why isn't the expression maintained?

Hauswirth: We don't find integration of AAV vectors in the retina for at least 6 months. There is an extrachomosomal persistence of double-stranded vectors that would be diluted in dividing cells. Also there is a slow reduction in the ocular levels of vectored PEDF beyond six months.

Kaleko: This may be an appropriate time to raise the issue of the effectiveness of immune privilege in the retina. My question is relevant to the use of the tetracycline-controlled gene expression system. These are foreign proteins. Additionally, if you are trying to treat null mutations by gene therapy, you will be putting a foreign protein into the retina. In some gene therapy circumstances this is problematic, for example, in the liver. Is there immune privilege in the back of the eye? Can we be comfortable going ahead with the Tet system even though it means the introduction of foreign proteins?

Friedlander: We may have deluded ourselves by using the term 'immune privilege': this is not an absolute in the back of the eye.

Bennett: There are limits to this immune privilege.

Hauswirth: In the diseased eye there is a significant amount of degeneration; I think under these circumstances the immune privilege breaks down.

Kaleko: Would you be comfortable using the Tet system in human eyes? It is a question we are asking ourselves at the moment, and I don't know the answer.

Hauswirth: I don't, either.

Friedlander: In eyes with neovascularization there is marked alteration of the normal blood–retinal barrier. If you think this is a problem in degenerating eyes it is even more so in neovascular eyes. These are big problems that will need dealing with at some point.

Hauswirth: A lot of the pathology is related to the leakiness of these vessels. The pathology is already there, in a sense. I guess it is related to the question Jean Bennett asked as to whether the RPE65 expression in the dog is releasing any specific RPE65 antibody. If it is, it doesn't seem to be a regionalizing antibody.

Bennett: There is no question that there is a humoral response when you put a foreign transgene in the retina. With AAV, at least when delivering the *RPE65* transgene, we haven't seen any evidence of a cellular immune response, which is good. This is what will be toxic.

Hauswirth: Different transgenes might elicit different responses.

Aguirre: One of the challenges is what happens when we have genes that are larger than the carrying capacity of the AAV vector. There doesn't seem to be anything on the horizon that will be able to work.

Hauswirth: Lentiviruses or adenovirus have a much larger capacity, but they have their own attendant sets of problems. There are alternatives. Jean Bennett has shown that a gene can be split and then recombined from two separate AAV vectors (Reich et al 2003).

Bolz: The question of large genes is especially relevant for Usher type 1. This is a severe problem because people do not only have vision problems but they are also deaf. Do you think genes like cadherin 23 (Bolz et al 2001, Bork et al 2001, Petit 2001) could be replaced in parts? Regarding possibilities to test the effects of a therapy approach in animals, we would still face the problem that the current mouse models have no retinal degeneration. Usher 3 (Joensuu et al 2001) is caused by a small gene and thus may be a better target for gene therapy, but again, we don't have an animal model.

Hauswirth: The Clarin 1 gene in Usher 3 does fit the AAV vector and we are in a consortium to do that. Of course, we need an animal model. The vector took a month to construct but the animal will take a year and a half. The FDA may not require an animal model for clinical trial approval if you can demonstrate that the disease is severe enough. Usher's may fall into this category. For cadherin 23 it would be nice to do the biochemistry on it: we may be able to trim it significantly. This will require some serious biochemical understanding.

Zack: PEDF is a nice system in which AAV and adenovirus have both been used. In this system, can you compare them? More specifically, it is clear that adenovirus without PEDF has an anti-angiogenic effect. It didn't look like AAV has as much effect. Why?

Hauswirth: The adenovirus data aren't mine, of course. I would say that adenovirus has a more profound inflammatory response in the eye. It is clearly a transient effect. Gutless vectors may last longer. Peter Campochiaro has done both side-by-side, one eye adenovirus and one eye AAV. Looking at the data, it seems that adenovirus-PEDF expression is slightly better.

LaVail: Presumably the goal here is to get in and then get out. Is that right?

Hauswirth: No, I think it is the other way round. The patients that will initially be treated will be ones with active choroidal NV (CNV). We will target those patients who are at high risk of getting a subsequent neovascular event, in which case we want expression to persist so that subsequent CNV can be prevented or minimized.

Kaleko: The data that have been generated so far on early-generation AV vectors have shown that they are extraordinarily effective at giving very high level expression from RPE. It is quite amazing. However, expression is self-limited — sometimes less than a month. With Peter Campochiaro, we have used gutless AV vectors but the dose has been fairly high. The gutless AV persisted longer than the early generation AV. However, after several months we start to see photoreceptor toxicity. On the other hand, Dr Kochanek has recently published a paper (Kreppel et al 2002) in which he used a much lower dose. The vector achieved six months of expression. I don't believe there was much toxicity. Do we believe that AV will give years of expression? I am not confident of this.

Hauswirth: There has to be a mechanism for persistence, for example efficient chromosomal integration, and as far as I am aware this is not common for AV.

Kaleko: For our first attempts, we have opted for the long-term expressing vectors rather than AV.

LaVail: How long have you seen expression?

Kaleko: We use a lentiviral vector and have watched expression for about 8 months now.

Hauswirth: We have seen expression with AAV for a rat's lifetime. I don't think this is the issue.

Kaleko: We had the choice of gutless AV or lentivirus, and we chose lentivirus because we were more confident in its ability to achieve sustained expression.

Gal: Coming back to the animal models for ROP, there are two human inherited diseases that are thought to be quite similar. First, Norrie disease. There was a paper showing that in a very small number of babies with ROP there are mutations in the Norrie gene (Shastry et al 1997). However, the data were not very convincing and have not yet been confirmed in an independent sample. Second, it has recently been announced that the gene for exudative vitreoretinopathy (EVR), which is the closest model for ROP, has been cloned (Robitaille et al 2002). It is *FZD4*, a homologue of the *Drosophila* frizzled. It would be interesting to include the corresponding animal models.

Bird: The disease is indistinguishable from ROP.

Hauswirth: So what you are suggesting is that there may be a better first target for AAV-vectored anti-neovascular therapy?

Bird: You were talking about treatment of AMD and CNV. It shouldn't be assumed that once CNV has been initiated that vision is lost and is unrescuable. That is actually not the case. It is the differential effect of your agent on pigment epithelium and the new vessel complex that is going to rescue vision. It is eminently rescuable in the early stages. In your photocoagulation model, in the untreated one there was a very nice RPE coverage of the new vessel complex on left hand side and a good looking photoreceptor cell population over that area. That is, the photoreceptor cells were not lost.

Hauswirth: I think you would agree that if we could prevent subsequent neovascular events this would be beneficial.

Bird: It is a matter of the predictability. If the risk is something like 6% per year, you have to express the therapeutic agent for a very long time before there is a therapeutic benefit evident. I would much rather treat for the people with early-stage CNV, because it is a rescuable situation. In occult CNV the vessels can exist for a long time before vision is lost. Providing the pigment epithelium survives intact then the vision remains good.

References

Bolz H, von Brederlow B, Ramirez A et al 2001 Mutation of CDH23, encoding a new member of the cadherin gene family, causes Usher syndrome type 1D. Nat Genet 27:108–112

Bork JM, Peters LM, Riazuddin S et al 2001 Usher syndrome 1D and nonsyndromic autosomal recessive deafness DFNB12 are caused by allelic mutations of the novel cadherin-like gene CDH23. Am J Hum Genet 68:26–37

Joensuu T, Hamalainen R, Yuan B et al 2001 Mutations in a novel gene with transmembrane domains underlie Usher syndrome type 3. Am J Hum Genet 69:673–684 [Erratum in Am J Hum Genet 69:1160]

Kreppel F, Luther TT, Semkova I, Schraermeyer U, Kochanek S 2002 Long-term transgene expression in the RPE after gene transfer with a high-capacity adenoviral vector. Invest Ophthalmol Vis Sci 43:1965–1970

Petit C 2001 Usher syndrome: from genetics to pathogenesis. Annu Rev Genomics Hum Genet 2:271–297

Reich S J, Auricchio A, Hildinger M et al 2003 Efficient trans-splicing in the retina expands the utility of adeno-associated virus as a vector for gene therapy. Human Gene Ther 14:37–44

Robitaille J, MacDonald MLE, Kaykas A et al 2002 Mutant frizzled-4 disrupts retinal angiogenesis in familial exudative vitreoretinopathy. Nat Genet 32:326–330

Shastry BS, Pendergast SD, Hartzer MK, Liu X, Trese MT 1997 Identification of missense mutations in the Norrie disease gene associated with advanced retinopathy of prematurity. Arch Ophthalmol 115:651–655

Gene therapy for Leber congenital amaurosis

Jean Bennett

FM Kirby Center for Molecular Ophthalmology, Scheie Eye Institute and Department of Ophthalmology, University of Pennsylvania, 310 Stellar-Chance Labs, 422 Curie Boulevard, Philadelphia PA 19104-2689, USA

Abstract. Recent success in delivering vision to a canine model of a severe, early-onset blinding disease, Leber congenital amaurosis (LCA) (Acland et al 2001) demonstrates that adeno-associated virus serotype 2 (AAV2) is capable of delivering a corrective gene to the target retinal cells. Results of these studies indicate long-term rescue of vision as assessed by psychophysical, behavioural and molecular biological studies. Preliminary results of studies in progress are described and the implications of these results with respect to developing human clinical trials for LCA and for other retinal degenerative diseases are discussed.

2004 Retinal dystrophies: functional genomics to gene therapy. Wiley, Chichester (Novartis Foundation Symposium 255) p 195–207

Progress in retinal gene therapy

Mutations in a large number of different genes have been identified and implicated in the pathophysiology of a variety of retinal degenerative diseases (Retnet, *http:// www.sph.uth.tmc.edu/RetNet*). With the concurrent identification of the homologous genes/mutations in animal models with retinal degeneration, it has become possible to test for therapeutic effects mediated by delivery of nucleic acids. Surgical techniques with which to deliver transgenes to target retinal cells have been developed. In addition, a number of different gene transfer reagents have been identified and many have been used to deliver nucleic acids efficiently to target retinal cells. With the development of methods for *in vivo* retinal gene delivery (reviewed by Bennett 2000, Bennett & Maguire 2000, Hauswirth & Beaufrer 2000, Dejneka & Bennett 2001) it has become possible to evaluate efficacy of gene therapy in animal models for particular forms of these blinding diseases. Successful gene-based intervention has been reported using a number of different vectors, a variety of animal models, a number of different disease targets, and using multiple different strategies. Rescue has been achieved with a number of different vectors, including adenovirus, adeno-associated virus (AAV), lentivirus and gutted adenovirus. Successful strategies include delivery of wild-type cDNA in the case of loss-of-function disease (Bennett et al 1996, Kumar-Singh & Farber 1998, Sarra et al 2001, Vollrath et al 2001, Ho et al 2002), delivery of ribozymes which specifically target the mutant mRNA in the case of gain-of-function disease (Lewin et al 1998, LaVail et al 2000), delivery of neurotrophic factors (Liang et al 2001a,b, McGee et al 2001) which preserve the integrity of the photoreceptors, and delivery of genes with anti-apoptotic function (Bennett et al 1998) in order to limit programmed cell death.

In the majority of studies, the therapeutic reagent is delivered to one eye and a control agent to the contralateral eye. Therapeutic endpoints have included electrophysiological evidence, biochemical measures, and evidence of histological rescue coupled with evidence that the transgene is expressed. Efficacy of treatment can be appreciated by comparing the outcome measures in the experimental and control eyes. Efficacy is most easily assessed in paradigms where the disease is severe and rapidly progressive. In many of these paradigms, rescue has been detected months after treatment.

Recently, a canine model was selected for study of efficacy of gene therapy for a particularly severe early-onset blinding disease, Leber congenital amaurosis. LCA is usually diagnosed in infancy as children with this disease are usually born with minimal visual function. In addition, abnormal ocular movements (nystagmus) are apparent early in life. LCA can be caused by mutations in at least half a dozen genes (Cremers et al 2002, *http://www.sph.uth.tmc.edu/RetNet*). *RPE65* is one such gene. It is named for the evolutionarily conserved retinal pigment epithelium (RPE)-specific 65 kDa protein which it encodes (Bavik et al 1992, Redmond & Hamel 2000). *RPE65* mutations account for a significant proportion of LCA: approximately 7–16% of these cases (Marlhens et al 1997, Morimura et al 1998, Lotery et al 2000). Mutations in *RPE65* can also cause early-onset retinitis pigmentosa (RP) (Gu et al 1997, Morimura et al 1998, Thompson et al 2000).

The *RPE65* mutant dog suffers from an autosomal recessive inheritance of a retinal disease with severe visual deficits. The mutation in this animal, a homozygous 4 bp deletion in the canine RPE65 coding sequence leading to a premature stop codon (Aguirre et al 1998, Veske et al 1999) is responsible for the phenotype (Narfstrom et al 1989, Wrigstad 1994, Aguirre et al 1998). There are abnormal qualitative and quantitative measures of visual function early in life. There is a relatively slow degenerative component, however. Fundus examination of dogs homozygous for the defect reveals prominent RPE inclusions and slightly abnormal rod photoreceptor morphology present within the first year of life (Wrigstad 1994). There is also a lack of immunohistochemically detectable RPE65 protein (Aguirre et al 1998). As the dogs age, a slowly progressive photoreceptor degeneration becomes apparent. Photoreceptor degenerative changes and loss are readily detectable in older (~5 year old) dogs (Wrigstad 1994). Biochemical studies reveal a lack of 11-*cis*-retinol and an accumulation of retinyl esters.

In addition to the *RPE65* mutant dog, there is a murine model available for studies of *RPE65*-based LCA. An *Rpe65* knockout (*Rpe65$^{-/-}$*) mouse was generated by T.M. Redmond and colleagues and this animal also suffers from an early-onset retinal degeneration (Redmond et al 1998, Redmond & Hamel 2000).

Initial results: rescue of vision in the *RPE65* mutant dog

The experimental objective was to deliver wild-type copies of the disease-causing gene (*RPE65*) to cell populations primarily affected by the gene mutation (i.e. the RPE). Affected dogs were identified through molecular diagnosis and by performing baseline electroretinograms (ERGs) (Acland et al 2001).

Unilateral subretinal injections of the corrective gene (the canine RPE65-encoding cDNA) carried by AAV serotype 2 (AAV2; AAV.RPE65) were performed, thereby exposing the diseased cells to the virus. Injections were performed when the animals were 4 months of age. A constitutive promoter (chicken β actin) was used and the amount of virus delivered was 1–5×10^{10} infective units of AAV. Approximately 20–30% of the retina was exposed to the virus. The contralateral eyes received control injections of intravitreal AAV.RPE65. Intravitreal delivery of AAV2 results in gene delivery to the inner retina but not the outer retina/RPE (Dudus et al 1999).

The animals were followed over time with regular clinical evaluations/laboratory testing to evaluate local/systemic toxicity. Evaluations included fundus exams, immune studies and periodic laboratory testing (measuring complete blood count [CBC] and blood chemistries, and also evaluations of the possibility that the virus/transgene escaped beyond the injection site). In eyes that received subretinal injections, retinas reattached within 24 h. There was minimal inflammation immediately after surgery. Injected portions of the retina could be reliably identified based on ophthalmoscopically detectable pigmentary changes at the level of the RPE (Acland et al 2001).

Animals appeared healthy and active after treatment and laboratory results showed that the CBC and blood chemistries were within normal limits through the course of the study by Acland et al (2001). There was no evidence of escape of virus from the eye. Immune studies showed antibody responses to both the vector and the transgene in sera and/or anterior chamber fluid (Acland et al 2001). There was no evidence that this had a negative effect on the animal or the eye itself.

Visual function testing was performed three to four months after treatment and included ERGs, pupillometry, measurement of visual evoked cortical potentials (VECP)s and measures of visual behaviour (Acland et al 2001).

In the *RPE65* mutant dog, rod photoreceptor-specific ERG waveforms are essentially flat from birth, even using high intensity stimuli. Quantitative testing of retinal function using ERGs two and a half to four months after treatment revealed waveforms similar to those present in wild-type dogs, although reduced in amplitude (Acland et al 2001).

VECPs were recorded in order to determine whether there was evidence of cortical vision. There were definitive VECP waveforms in the *RPE65* mutant dogs after subretinal delivery of AAV.RPE65 (Acland et al 2001).

There was evidence of visual behaviour in treated *RPE65* mutant dogs soon after treatment. Those animals could navigate an obstacle course by four months after treatment using their experimentally treated eyes (Acland et al 2001). The animals also demonstrated preferential gaze by the treated eye (Bennett 2003).

Studies in progress

The dogs treated in the initial studies nearly two and a half years ago, continue to be monitored for therapeutic efficacy and for toxicity. There does not appear to be any significant diminution of visual function as compared to the measures made two and a half months after the treatment. Pupillometry and ERG responses were similar at the two and a half month and the 13 month time points (Acland et al 2002). The dogs appear to be healthy and comfortable.

Studies in progress aim to further characterize the therapeutic effects, to document the stability of these effects and to determine the age limits of treatment. In additional treated dogs, therapy is still effective even when administered to dogs older than four months (Acland et al 2002). In addition, preliminary studies evaluating the effects of the treatment on ocular motility indicate that nystagmus is significantly reduced in eyes treated with successful subretinal injection of AAV.RPE65 (Jacobs et al 2003).

Studies, in progress, in the *Rpe65$^{-/-}$* mouse, indicate a rescue of visual function in eyes that receive adequate amounts of RPE65 protein through delivery of either the canine or the human *RPE65* cDNA (N. S. Dejneka, E. M. Surace, T. S. Aleman, A. V. Cideciyan, A. M. Maguire, S. G. Jacobson, J. Bennett, unpublished data).

From bench to bedside: what are the necessary steps before embarking on a human clinical trial for LCA?

A recombinant AAV, generated using AAV serotype 2 (AAV2) capsids, was used in the initial studies involving the RPE65 mutant dog, to deliver a wild-type version of the canine *RPE65* cDNA. This rescued visual function as assessed by psychophysical and behavioural studies. While this information demonstrates that AAV2 is sufficient for rescuing the phenotype, it is important to identify the optimal set of conditions for rescue before embarking on a clinical trial. There are many variables to consider, including selection of the optimal vector, incorporation of cell-specific regulatory elements, extent of the retina that should be transduced, ability to readminister the vector, and age of treatment. The recent demonstration that one can control retinal cellular specificity, onset and levels of transgene expression by packaging the AAV in capsids from AAVs of different serotypes may be relevant (Auricchio et al 2001). It is imperative that the vector/transgene delivery be safe with respect to both the eye and to the rest of the body. Special attention must be paid to the possibility that the transgene could potentially be introduced to the brain via unwanted ganglion cell/optic nerve targeting (Dudus et al 1999). In addition, ideally, the transgene will be maintained long-term in the target cells and will not target/disrupt any other genes/regulatory elements. Such a phenomenon could result in uncontrolled cell division as appears to be the case in a child treated with a retrovirus for severe combined immune deficiency (SCID) (Stolberg 2002).

In addition to gene delivery issues (and once the limitations of rescue in animals have been delineated), human subjects who are eligible for treatment must be identified. Mutation studies must verify the source of the disease; non-disease-causing mutations should be excluded. The disease pathogenesis/time-course in humans should be characterized. Outcome measures should be developed for human testing which will provide a suitable and non-invasive means of evaluating the success of the treatment. An exit strategy should be developed in case there are any unforeseen toxicities. One such strategy could be to remove the transduced cells via photocoagulation. Such a strategy would need to be tested first in animal models.

Can the success in the canine LCA gene therapy studies be extrapolated to other forms of retinal degeneration?

A major challenge is how to extrapolate the successful treatment of the *RPE65* disease to successful treatment of other severe retinal degenerative diseases. What are the factors which led to success in the *RPE65* mutant dog? The disease affecting the *RPE65* mutant dog has a relatively slow degenerative component. Clearly, the cells need to be present and viable for a long enough period for them to be rescued. The therapeutic window during which the retina can be rescued should be considered carefully for each disease under scrutiny. It may be that some diseases occur so rapidly, that treatment will have to be administered early in life (or even during gestation; Surace et al 2003).

Vector selection is likely to play a large role in the ability to slow disease progression. It is important for the vector to target the appropriate cell types in an efficient and stable manner. It is also important that transgene expression peaks within the therapeutic window. AAV2, which has a slow onset of transgene expression (Bennett et al 1998, 1999, 2000), may not lead to high enough transgene levels at the appropriate time to slow the disease progression of rapidly progressive degenerative diseases. Instead, it might be more effective to employ a viral vector which results in a rapid onset of transgene expression.

Finally, there may be requirements of the disease model that will have to be met by developing alternative vectors/vector strategies. There are certain transgene cassettes, for example, which are too large for the packaging limitations imposed by AAV. In order to use AAV, it will be necessary to split the cassette into pieces, each within the limit on packaging size of AAV (4.8 kb) and package the pieces into separate AAVs. The pieces should be selected so that the cell can splice them together (i.e. 'trans-splicing'). Such a strategy has been demonstrated to function efficiently in retinal cells (Reich et al 2003). Alternatively, one could select a non-AAV vector which is capable of carrying the transgene cassette in question. Lentivirus, for example, can carry a cassette of ~ 8 kb, and gutted adenovirus, can carry a cassette of > 30 kb.

In summary, AAV2 was used to deliver corrective genes subretinally to a canine model of a severe retinal disease and efficacy of treatment was readily apparent. Continued success in this paradigm may lead to identification of critical parameters for the success/safety of gene therapy for this and other retinal degenerative diseases. Success in animal models of the diseases will be used ultimately to design human clinical trials.

Acknowledgements

The work that is described in this report was performed by Acland et al (Acland et al 2001) and through additional collaborative studies including L.F. Dell'Osso, J. Jacobs, R. Hertle, N.S. Dejneka, and E.M Surace, E.M. Stone and K. Palczewski. The expert technical assistance from G. Antonini, N. Bennett (and Bitmax, Inc), and A. Nickle is gratefully acknowledged. Support was from NIH grants EY10820, EY11123, NS36202, EY06855, EY11142, EY13132 and U10EY013729; The Foundation Fighting Blindness; Research to Prevent Blindness; T.L. Andresen Endowment, the Macular Vision Research Foundation, the LIFE Foundation, the Steinbach Foundation, the Mackall Foundation Trust, and the F. M. Kirby Foundation.

References

Acland GM, Aguirre G, Maguire AM et al 2001 Gene therapy restores vision in a canine model of childhood blindness. Nat Genet 28:92–95
Acland GM, Aguirre G, Aleman T et al 2002 Continuing evaluation of gene therapy in the rpe65 mutant dog. Invest Ophthalmol Vis Sci (abstr 4593)

Aguirre G, Baldwin V, Pearce-Kelling S, Narfstrom K, Ray K, Acland GM 1998 Congenital stationary night blindness in the dog: common mutation in the RPE65 gene indicates founder effect. Mol Vis 4:23

Auricchio A, Kobinger G, Anand V et al 2001 Exchange of surface proteins impacts on viral vector cellular specificity and transduction characteristics: the retina as a model. Hum Mol Genet 10:3075–3081

Bavik C, Busch C, Eriksson U 1992 Characterization of a plasma retinol-binding protein membrane-receptor expressed in the retinal pigment epithelium. J Biol Chem 267:23035–23042

Bennett J 2000 Gene therapy for retinitis pigmentosa. Curr Opin Med Ther 2:420–425

Bennett J 2003 Gene therapy for childhood onset blindness. In: Templeton NS (ed) Gene therapy: therapeutic mechanisms and strategies, 2nd edn, Marcel Dekker Inc, New York, in press

Bennett J, Maguire AM 2000 Gene therapy for ocular disease. Mol Ther 1:501–505

Bennett J, Tanabe T, Sun D et al 1996 Photoreceptor cell rescue in retinal degeneration (*rd*) mice by in vivo gene therapy. Nat Med 2:649–654

Bennett J, Zeng Y, Bajwa R, Klatt L, Li Y, Maguire AM 1998 Adenovirus-mediated delivery of rhodopsin-promoted *bcl-2* results in a delay in photoreceptor cell death in the *rd/rd* mouse. Gene Ther 5:1156–1164

Bennett J, Maguire AM, Cideciyan AV et al 1999 Stable transgene expression in rod photoreceptors after recombinant adeno-associated virus-mediated gene transfer to monkey retina. Proc Natl Acad Sci USA 96:9920–9925

Bennett J, Anand V, Acland GM, Maguire AM 2000 Cross-species comparison of in vivo reporter gene expression after recombinant adeno-associated virus-mediated retinal transduction. Methods Enzymol 316:777–789

Cremers F, van den Hurk J, den Hollander A 2002 Molecular genetics of Leber congenital amaurosis. Hum Mol Genet 11:1169–1176

Dejneka NS, Bennett J 2001 Gene therapy and retinitis pigmentosa: advances and future challenges. Bioessays 23:662–668

Dudus L, Anand V, Acland GM et al 1999 Persistent transgene product in retina, optic nerve and brain after intraocular injection of rAAV. Vision Res 39:2545–2554

Gu S-M, Thompson DA, Srikumari CRS et al 1997 Mutations in *RPE65* cause autosomal recessive childhood-onset severe retinal dystrophy. Nat Genet 17:194–197

Hauswirth W, Beaufrer L 2000 Ocular gene therapy: quo vadis? Invest Ophthalmol Vis Sci 41:2821–2826

Ho TT, Maguire AM, Aguirre GD et al 2002 Phenotypic rescue after adeno-associated virus-mediated delivery of 4-sulfatase to the retinal pigment epithelium of feline mucopolysaccharidosis VI. J Gene Med 4:613–621

Jacobs JB, Dell'Osso LF, Hertle RW, Bennett J, Acland G 2003 Gene therapy to abolish congenital nystagmus in RPE65-deficient canines. Invest Ophthalmol Vis Sci (abstr 4249)

Kumar-Singh R, Farber DB 1998 Encapsidated adenovirus mini-chromosome-mediated delivery of genes to the retina: application to the rescue of photoreceptor degeneration. Hum Mol Genet 7:1893–1900

LaVail MM, Yasumura D, Matthes MT et al 2000 Ribozyme rescue of photoreceptor cells in P23H transgenic rats: long-term survival and late-stage therapy. Proc Natl Acad Sci USA 97:11488–11493

Lewin AS, Drenser KA, Hauswirth WW et al 1998 Ribozyme rescue of photoreceptor cells in a transgenic rat model of autosomal dominant retinitis pigmentosa. Nat Med 4:967–971

Liang F-Q, Aleman TS, Dejneka NS et al 2001a Long-term protection of retinal structure but not function using rAAV.CNTF in animal models of retinitis pigmentosa. Mol Ther 4:461–472

Liang F-Q, Dejneka NS, Cohen DR et al 2001b AAV-mediated delivery of ciliary neurotrophic factor prolongs photoreceptor survival in the rhodopsin knockout mouse. Mol Ther 3:241–248

Lotery A, Namperumalsamy P, Jacobson S et al 2000 Mutation analysis of 3 genes in patients with Leber congenital amaurosis. Arch Ophthalmol 118:538–543

Marlhens F, Bareil C, Griffoin J-M et al 1997 Mutations in RPE65 cause Leber's congenital amaurosis. Nat Genet 17:139–141

McGee Sanftner L, Abel H, Hauswirth WW, Flannery JG 2001 Glial cell line derived neurotrophic factor delays photoreceptor degeneration in a transgenic rat model of retinitis pigmentosa. Mol Ther 4:622–629

Morimura H, Fishman GA, Grover SA, Fulton AB, Berson EL, Dryja TP 1998 Mutations in the RPE65 gene in patients with autosomal recessive retinitis pigmentosa or Leber congenital amaurosis. Proc Natl Acad Sci USA 95:3088–3093

Narfstrom K, Wrigstad A, Nilsson SE 1989 The Briard dog: a new animal model of congenital stationary night blindness. Brit J Ophthalmol. 73:750–756

Redmond TM, Hamel CP 2000 Genetic analysis of RPE65: from human disease to mouse model. Methods Enzymol 317:705–724

Redmond TM, Yu S, Lee E et al 1998 Rpe65 is necessary for production of 11-cis-vitamin A in the retinal visual cycle. Nat Genet 20:344–351

Reich SJ, Maguire A, Auricchio A et al 2003 Trans-splicing AAV vector system expands the packaging capacity of AAV gene therapy vectors for delivery of large transgenes to the retina. Hum Gene Ther 14:37–44

Sarra G-M, Stephens C, de Alwis M et al 2001 Gene replacement therapy in the retinal degeneration slow (rds) mouse: the effect on retinal degeneration following partial transduction of the retina. Hum Mol Genet 10:2353–2361

Stolberg SG 2002 Trials are halted on a gene therapy. New York Times

Surace E, Auricchio A, Reich S et al 2003 Delivery of adeno-associated viral vectors to the fetal retina: impact of viral capsid proteins on retinal neuronal progenitor transduction. J Virol 77:7957–7963

Thompson D, Gyurus P, Fleischer L et al 2000 Genetics and phenotypes of RPE65 mutations in inherited retinal degeneration. Invest Ophthalmol Vis Sci 41:4293–4299

Veske A, Nilsson SE, Narfstrom K, Gal A 1999 Retinal dystrophy of Swedish briard/briard-beagle dogs is due to a 4-bp deletion in RPE65. Genomics 57:57–61

Vollrath D, Feng W, Duncan J et al 2001 Correction of the retinal dystrophy phenotype of the RCS rat by viral gene transfer of Mertk. Proc Natl Acad Sci USA 98: 12584–12589

Wrigstad A 1994 Hereditary dystrophy of the retina and the retinal pigment epithelium in a strain of briard dogs: a clinical, morphological and electrophysiological study. PhD Thesis, Linkoping University Medical Dissertations, Sweden

DISCUSSION

Kaleko: I am intrigued by the fact that you can treat older dogs and older mice. It was my understanding that once amblyopia occurs, it is relatively refractory to treatment. Do you think that one of the differences might be that when a human gets amblyopia, the brain has purposely suppressed vision from one eye, whereas in your animal models, there has been no active suppression of visual pathways?

Bennett: There may be just enough light getting into the eye in these young animals so that the circuitry is established. At the highest intensity of light there is a tiny blip in the electroretinograms. Perhaps enough is getting in to set up the circuitry.

Bird: Amblyopia mostly addresses form vision, rather than light perception. An ERG is not going to measure that, unless you use edge detection.

Zack: In classical amblyopia children have some vision. Related to this, in some of the dogs and mice that have been sacrificed, has anyone looked at the wiring through the visual cortex?

Aguirre: The short answer is no. We are planning to do functional imaging in these dogs before and during the therapy. One of the questions we would like to address is whether we have an expansion of the cortex that is subserving vision. The more visual these dogs are, the more their abilities improve. This may just be an adaptation effect without any changes in the wiring of the cortex. But it may be taking place, as well, at a cortical level. It will be easier to see this in a non-invasive temporal manner.

Friedlander: I am intrigued by your comments about the success of gene therapy and its relationship to the level of transgene expression. I was struck by one of your earlier papers in which you looked at the *Rd/Rd* mouse. It was striking how little cGMP phosphodiesterase was made in the cases where there was phenotypic rescue. Do we expect to see lots of differences between different diseases? If we only need to get a small amount of gene product expressed, perhaps our concerns about over expression can be minimized by giving minimal doses as opposed to what most of us try to do, which is to get maximal dosing. How do you quantify how much RPE65 is made in these dogs?

Bennett: That is an interesting question. This is work in progress. We all started out in the gene therapy field thinking that more is better. This is how we first saw the therapeutic effects. Now the trend is to back-off, because more may actually be toxic, and we may not need more. One point I need to make is that we are only treating a portion of the retina. If you look at the scale bars on the ERGs you will see that the amplitudes of the treated eyes are proportional to the amount of retina that was transduced. They are therefore smaller in the treated eye than you would see in a wild-type eye. This may be sufficient for vision. If you were just aiming to treat the macula and restore central vision then perhaps you wouldn't worry about treating the whole retina anyway.

Friedlander: Getting back to the issue of distribution and delivery, I am talking about absolute amounts. We see this even in non-gene therapy trials, such as the anti-angiogenic compounds we work with. When we do dose–response studies we sometimes will see a maximal response at a lower dosing.

LaVail: On the issue of delivery, I am not sure why you only transduce such a small area of the mouse retina. When we inject in rat retinas, we can show the GFP and also rescue across the whole retina.

Bennett: We have been able to achieve this by injecting a small volume so as to purposefully treat only a portion of the retina. We wanted to use some of the retina as its own control by leaving it untreated.

Aguirre: Large volume injections are problematic in the dog. They have a very well formed vitreous, so there is an upper limit of about 175–200 μl that can be injected. If there is no reflux out of the needle track, this represents about a third of the maximal dose.

Hauswirth: We are in a fortunate position that we can now back up on the dose. We now have to come to grips with a dose–response in dogs that is not going to be easy with the small number of dogs available.

Bennett: We have done experiments in dogs where we keep the same dose but inject a larger volume of solution into the subretinal space so that we treat more of the retina. There you have to do vitrectomy just to allow for the volume. This does seem to result in the same sort of response.

Kaleko: Dose–response curves can have a downside at high doses. We may lose effectiveness if we put in too much.

Aguirre: I wanted to comment on the issue raised by Marty Friedlander on whether all the cells are treated. This disease has an advantage. Unlike a photoreceptor cell, which must be treated directly, the end product of this treatment is vitamin A in the form that will get into the inter-photoreceptor space and diffuse. Thus you don't need to treat every RPE cell; it is enough just to hit many of them because the product is able to diffuse to adjacent cells.

Friedlander: What this may be telling us is that every disease may be different. Depending on the gene product that is targeted, what may work in one situation may not work in others.

Swaroop: You briefly mentioned the issue of virus integration in the genome. My understanding is that there are a large number of defined viral integration sites in the genome. Do you think theoretically it will be possible to design a vector that can go to certain target sites away from the regions that are oncogenic?

Hauswirth: The current understanding of AAV integration is that there is no subset of preferred sites. Secondly, with the viral *Rep* gene, if you put it back in you get nearly 100% integration into one site in chromosome 19.

Swaroop: That may be true for AAV, but there are many other viruses for which there are preferred sites in the genome. It is possible that if you have other vectors, you might be able to design targeting sequences for AAV.

Bennett: I am aware of one study which does what you are suggesting, by Michelle Calos at Stanford (Olivares et al 2002, Ortiz-Urda et al 2002). She has used this for therapy in a haemophilia model in mice and in epidermolysis bullosa. She has harnessed the bacteriophage integration sequences and has found that mice and other mammals have homologous sequences. She is able to use these to target to specific sites.

Zack: To change the subject a little, Gus Aguirre and colleagues have identified a rhodopsin mutant dog. Gus, can you tell us more about this?

Aguirre: We published this in April 2002 (Kijas et al 2002), and it involves a T4R mutation. The retina develops normally both in the heterozygous and in the homozygous mutant animals. Rhodopsin is targeted to the photoreceptor outer segments. It affects a glycosylation site but does not seem to affect the targeting. Rhodopsin is properly located. There is a specific functional defect similar to what Sam Jacobson has described in the type IIB human rhodopsin mutations: with an intense bleaching light, the recovery of the photoresponse is delayed. We have only looked at this so far up to 40 minutes. This same functional defect is seen in the Pro23His and T17M mutations in human. This is an endpoint that is measurable. We don't have to wait for the disease to develop but we are able to look to see whether we can restore the rhodopsin bleaching and regeneration properties to normal in animals treated with gene therapy. This suggests that the problem is not with the bleaching but with a post transduction step. The young animals have a normal ERG. What is nice is that this is a slow disease with an early detectable phenotype that is applicable to human patients with type 2 mutations in rhodopsin.

Zack: What gene transfer approaches are you planning?

Aguirre: Bill Hauswirth and Al Lewin are working on a ribozyme.

Hauswirth: The kinetics are not great so we are going back and trying to fine-tune it.

Gal: The human disease is regional. Is this the case for the dog?

Aguirre: Yes, when Sam Jacobson looks at the dogs with OCT, he initially see a patchy degeneration, which eventually becomes widespread. Someone in the Netherlands described a T4 mutation that at one time was believed to be a cone–rod dystrophy. Fortunately the insert size needed for gene therapy of rhodopsin mutations is suitable for use with AAV vectors. This brings me back to the question we discussed earlier about the carrying capacity of viral vectors. The problem we face with *RPGR* is that the gene is 5.2 kb, which doesn't fit in AAV. We would like to do one injection, and we don't have enough biochemical information about the protein to be able to chop it up. In the dog AV is a real problem, and lentivirus mainly targets the RPE. *ABCR* is another large gene which conceivably will have the same delivery problems.

Hauswirth: If you really need AAVs and you need to put in a cDNA that won't package because of its size, currently there is only one option. John Engelhardt has developed a recombination system where if you find an intron that splits the gene you can use those sequences to recombine two parts of the cDNA delivered in two AAV vectors at somewhere around the 5% level and express a gene product in the cell that cannot be packaged in a single vector. The problem is the low efficiency, but this may not be a problem when using high multiplicity vectors in the retina.

Bennett: We have a paper in press describing the use of such a system for the delivery of a split gene to the retina. By manipulating the capsid serotype we can get very efficient gene transfer of the two different AAVs which combine beautifully in the photoreceptors and retinal pigment epithelium (Reich et al 2003). This may be the way to go for RPGR.

Travis: In the T4R dog, if you go to very low intensity stimulation with ERG, do you see reduced sensitivity in the dark-adapted animal?

Aguirre: You don't notice a recovery defect using single-flash recordings because you are not bleaching enough of the photopigment. If you go bleach 40% of the photopigment you get normal recovery. However, total bleaches result in a prolonged recovery of the rod photoresponse to pre-bleach levels.

Travis: They just have slow recovery after a bleach, then.

Aguirre: You don't notice a recovery using single-flash recordings because you are not bleaching enough of the photopigment. If you go to 40% bleaching you will get normal recovery.

Travis: I would expect the T4R to be a non-glycosylated protein. From the description of delayed recovery, however, you would expect constitutive activation, but this seems more like it would be misfolded or mislocalized.

Aguirre: We were thinking this, but the immunostaining is very clear and shows that it is localized in the right site.

Hauswirth: Dr Shalesh Kaushel has been studying misfolded opsin proteins. If some fraction of the mutant is misfolded it is actually targeted to the ubiquitin disposal pathways. So you may not see a lot of protein mislocalized, but it may still be interfering with protein trafficking.

Travis: If there is a significant fraction of rhodopsin not in the right place, this would affect the threshold.

Aguirre: I would agree. In human patients it seems that under different environmental conditions the same mutation may have differing effects. One advantage of the dog for these studies is that, although our dogs are genetically quite heterogeneous, the diseases in our colony have a single mutant chromosome, and dogs share the same environment and diet. This allows one to control for these factors while examining phenotypic variability.

References

Kijas JW, Cideciyan AV, Aleman TS et al 2002 Naturally occurring rhodopsin mutation in the dog causes retinal dysfunction and degeneration mimicking human dominant retinitis pigmentosa. Proc Natl Acad Sci USA 99:6328–6333

Olivares EC, Hollis RP, Chalberg TW, Meuse L, Kay MA, Calos MP 2002 Site-specific genomic integration produces therapeutic factor IX levels in mice. Nat Biotechnol 20:1124–1128

Ortiz-Urda S, Thyagarajan B, Keene DR et al 2002 Stable nonviral genetic correction of inherited human skin disease. Nat Med 8:1166–1170 [Erratum in Nat Med 9:237]

Reich S J, Auricchio A, Hildinger M et al 2003 Efficient trans-splicing in the retina expands the utility of adeno-associated virus as a vector for gene therapy. Hum Gene Ther 14:37–44

Index of contributors

Non-participating co-authors are indicated by asterisks. Entries in bold indicate papers; other entries refer to discussion contributions.

Subject index

Page numbers in *italics* indicate tables.